Green Energy and Technology

Climate change, environmental impact and the limited natural resources urge scientific research and novel technical solutions. The monograph series Green Energy and Technology serves as a publishing platform for scientific and technological approaches to "green"—i.e. environmentally friendly and sustainable—technologies. While a focus lies on energy and power supply, it also covers "green" solutions in industrial engineering and engineering design. Green Energy and Technology addresses researchers, advanced students, technical consultants as well as decision makers in industries and politics. Hence, the level of presentation spans from instructional to highly technical.

Indexed in Scopus.

Indexed in Ei Compendex.

Mohammed El Amine Abdelli ·
Muhammad Shahbaz
Editors

Digital Economy, Energy and Sustainability

Opportunities and Challenges

Springer

Editors
Mohammed El Amine Abdelli
Western Economics and Management Laboratory-LEGO
University of Western Brittany
Brest, France

University of Salamanca
Salamanca, Spain

Muhammad Shahbaz
Department of Land Economy, Center for Energy and Environmental Policy Research
University of Cambridge
Cambridge, UK

School of Management and Economics
Beijing Institute of Technology
Beijing, China

ISSN 1865-3529 ISSN 1865-3537 (electronic)
Green Energy and Technology
ISBN 978-3-031-22381-5 ISBN 978-3-031-22382-2 (eBook)
https://doi.org/10.1007/978-3-031-22382-2

This Springer imprint is published by the registered company Springer Nature Switzerland AG
The registered company address is: Gewerbestrasse 11, 6330 Cham, Switzerland

Preface

The book is titled *Digital Economy, Energy and Sustainability: Opportunities and Challenges* and is part of the "Green Energy and Technology" book series.

According to the OECD, the digital economy is an umbrella term used to represent economies that focus on and often involve transferring information commodities or services through electronic commerce. Nowadays, the Internet serves as a key facilitator for exchanging sustainability knowledge. However, this kind of facilitation is difficult by the vast information available nowadays. The information systems that can quickly and effectively offer trustworthy evaluations of the information published on sustainability are missing in the digital economy. Also, an enormous energy crisis that will affect the entire world is imminent, and in the global energy environment, we live in a time where environmental protection efforts using advanced technology are well underway. This book focuses on Digital Economy, Energy and Sustainability issues to investigate trends and analyze environmentally impacting variables. In areas including transportation, green buildings, smart homes and smart cities, green AI and platforms of green enterprises utilizing computing models and methodologies are desperately needed. IoT and other digital tools potentials can offer answers where clever recent methods and strategies might be suggested. This book primarily investigates these methods to improve energy systems and focuses on improving energy consumption and employing alternative energy sources to safeguard the environment through methods, models, architectures and codes and their outcomes. An important aspect of the book is that it provides case studies that back up the theory of Digital Economy, Energy and Sustainability issues.

Brest, France — Mohammed El Amine Abdelli
Cambridge, UK/Beijing, China — Muhammad Shahbaz

Introduction

The book looks at how the digital economy may help the energy business meet its sustainability goals. It argues that various elements are accountable for and contribute to the many forms of sustainability of digital information services. Open Internet and data access are required to allow any human being to learn in such cases. Following a double-blind peer review, chapters highlighting the key opportunities and challenges of the Digital Economy, Energy and Sustainability were chosen. These chapters represent a research field pertinent to academics, practitioners and organizations. Two distinct themes are used to categorize the chapters. Six chapters of first part are devoted to the digital economy and sustainability. Second part has three chapters centered on development and energy processes.

The chapter's authors indicated that growth in the services sector has a bigger impact on alleviating poverty than growth in the manufacturing or agricultural sectors. This is plausible, considering how heavily these industries rely on non-renewable energy sources. Extreme weather is a result of rising carbon emissions, which also cause property damage to houses and businesses. However, the researchers of Chapter "Financial Inclusion and Environmental Sustainability" noted that the link between financial inclusion and environmental sustainability is examined in the study using Pearson correlation analysis. Programs and initiatives for financial inclusion in non-EU nations support environmental sustainability efforts. The findings also show a large and unfavorable relationship between environmental policy rigor and multifactor productivity growth that is environmentally adjusted. Chapter "eNaira Central Bank Digital Currency (CBDC) for Financial Inclusion in Nigeria" shows that observers from all over the world, including central banks, have expressed great interest in the Nigerian government's launch of the eNaira CBDC. This study investigates the possibilities for greater financial inclusion in Nigeria and considers the potential economic and banking sector impacts.

Chapter "Catalyzing Climate Finance for Climate Actions in MENA Countries: A Holistic View of Egypt, Morocco, and Tunisia" investigates that one of these difficult problems is the need for quick action on climate change. Making public and private climate finance affordable and accessible is hampered by several factors. It would be possible for nations to finance the investments they require to combat climate change if a cooperative effort involving all actors and integrating the mobilization of all resources was made. Chapter "Analysis of the Indicators of Environmental Performance in Algeria" noted that the energy market in Algeria is dominated by hydrocarbons, which generate 93% of export income. Since 2014, the trade balance finances have worsened due to the drop in oil prices. The transition to sustainable resource extraction has become a top priority for Algeria. Chapter "Circular Economy in Algeria: Strategies and Obstacles" explains that the actual trends toward a circular economy seek to achieve several successes favorable to sustainable development. Most nations aim to transition to this new model. However, developing nations like Algeria will find this process more difficult. In this paper, the authors discussed the various circular economy solutions and listed Algeria's difficulties that must be overcome to develop the sustainability norms that will pave the way for the shift to a circular economy. Chapter "Commercializing Bee Pollination to Increase Maize Productivity and Farmers' Economic Gains in Tanzania" noted that to raise maize production and economic gains to accomplish sustainability among smallholder farmers in Tanzania, this paper investigates the commercialization of bee pollination.

Chapter "Decomposing Energy-Related CO_2 Emissions in Tunisia Using the LMDI Approach" noted that the primary driver of CO_2 emissions growth in Tunisia is the country's GDP expansion, while energy intensity reductions and structural effects also help to lower CO_2 emissions. For the nation to achieve its planned 1.5–2 °C global warming levels, a greater understanding of energy-induced carbon emissions across all sectors would have significant policy implications. Finally, Chapter "Soil Properties and Carbon Sequestration Using Biochar and Compost on an Alfisol in Southwest Nigeria" noted that an Alfisol was the subject of a field experiment that lasted five years in southwest Nigeria. This study aimed to evaluate the impact of biochar and compost on soil characteristics, maize yield and carbon sequestration over a similar time frame. The authors concluded that the carbon captured improved the soil's characteristics, maize yield and accumulation.

Brest, France Mohammed El Amine Abdelli
Cambridge, UK/Beijing, China Muhammad Shahbaz

Contents

Digital Economy and Sustainability

Channeling Growth for Sustainable Development in Sub-Saharan Africa: Do Sectoral Patterns of Growth Matter?

Olaoye Olumide Olusegun

1 Introduction

The first of the 17 sustainable development goals (SDGs) is to end poverty by 2030 and give everyone in the world a chance to prosper and live a productive and rich life. Since the 1990s, eradicating poverty has become a central paradigm in development economics, championed by multilateral institutions, i.e., the World Bank and the International Monetary Fund (IMF). To end poverty, several scholars have noted that countries must spur growth to achieve this very important objective. Similarly, the early theory of economic development stresses that economic growth can help to reduce the level of poverty.

Many African countries have enjoyed strong economic growth over the last few decades, especially in the last 10–15 years (see Olaoye 2022a). Unfortunately, despite the increase in economic growth, the macroeconomic performance indices are poor and alarming. For example, although growth rate improved in 16 major sub-Saharan African countries (corresponding to nearly 75% of the total population), these countries are yet to channel the benefits of economic growth into the pockets of the poor (Arndt et al. 2016). Available evidence also shows that 18 out of the 20 countries with the highest shares of the world's extremely poor are in sub-Saharan Africa, while nearly half of the region's population lives in extreme poverty (IDS 2020).

Consequently, some scholars have argued that there is a need for the structural transformation[1] of the African economy (see Danquah et al. 2022). In this regard, some authors (see Ferreira et al. 2010; Adejumo et al. 2020) have argued that the

[1] Danquah et al. (2022) defined structural transformation as the shift from an agrarian economy to a more industrialized economy as well as the redistribution of income to the poor households. The authors concluded that structural transformation is crucial to inclusive economic growth and poverty reduction through creation of jobs and improving labor productivity.

O. O. Olusegun (✉)
Thomas Adewunmi University, Oko, Nigeria
e-mail: olumide.olaoye@tau.edu.ng

M. El Amine Abdelli and M. Shahbaz (eds.), *Digital Economy, Energy and Sustainability*, Green Energy and Technology, https://doi.org/10.1007/978-3-031-22382-2_1

services sector (this includes the information and telecommunication sector, the finance sector, media and entertainment industry, the healthcare industry, etc.) as the sector that can enhance green and quality growth which can significantly help to reduce poverty. According to them, the manufacturing and agricultural sectors, particularly in developing countries, are characterized by the excessive utilization of non-renewable sources of energy (such as wood fuels, petroleum, coal, and biofuel) in their operations, thereby increasing carbon emissions. The increase in carbon emissions creates extreme weather, which in turn damages homes and businesses, and this can prevent the poor from escaping poverty, and it is often the trigger that tips the vulnerable into poverty (Baloch et al. 2020; Khan 2019; The World Bank 2015).

2 Economic Growth and Poverty Reduction: A Brief Literature Review

A preponderance of empirical literature has examined the link between economic growth and poverty reduction (see Erlando et al. 2020; Adeleye et al. 2020; Breunig and Majeed 2019; Fosu 2017; Hailu and Nagaraja 2017; Moore and Donaldson 2016; Kudebayeva and Barrientos 2013; Mulok et al. 2012; Ferreira et al. 2010; Donaldson 2008; Ravallion 2002). In an early study, Ravallion (2002) found that poverty has an adverse effect on consumption growth and reduces the impact of economic growth on poverty reduction. In contrast, Fosu and Gafa (2020) found that income growth drives down poverty in most countries; however, income inequality played a crucial role in poverty behavior in many countries, indicating that with income re-distribution, a lot more progress would be made. Adeleye et al. (2020) find that economic growth exhibits poverty reduction properties. However, their result shows that inequality intensifies poverty and that inequality reduces the impact of growth on poverty reduction.

In contributing to the debate, Hailu and Nagaraja (2017) examined the impact of growth and inequality on poverty in the Amhara region of Ethiopia. The study found that increase in real household per capita expenditure results in a fall in the inequality gap and reduces the incidence of poverty in the Amhara region.

In some early studies, Dollar and Kraay (2000, 2001, 2002), using data from over 70 countries, found that an increase in real gross domestic product (real GDP) per capita leads to a significant decline in poverty. In another study, Agrawal (2008) found that the higher the growth rates, the faster the decline in poverty and the higher the real wages. Several other scholars have also supported the importance of economic growth for poverty reduction (see Deaton and Dreze 2001; Bhagwati 2001; Datt and Ravallion 2002). Ferreira et al. (2010) found that the effectiveness of economic growth in reducing poverty differs across sectors, space, and time. The authors showed that growth in the services sector substantially reduces poverty more than in agriculture or industry. The study concluded that economic growth did not significantly influence Brazil's poverty reduction efforts between 1985 and 2004.

While there have been a few attempts (Adeleye et al. 2020; Fosu 2017; Hailu and Nagaraja 2017) to evaluate whether the economic growth recorded in African countries in the last few decades has led to a reduction in the level of poverty, the empirical evidence presented thus far is limited.

Against this background, some fundamental questions arise. One, has economic growth translated to poverty reduction in sub-Saharan African countries? Others are: is the effectiveness of growth to reduce poverty dependent on the growth in different sectors of the economy? And how do sub-Saharan African countries achieve sustainable growth, reduce income inequality, and eradicate the high incidence of poverty across the region?

This study contributes to economic growth and poverty literature in the following significant ways:

One, the study investigates the quality of growth in SSA. Specifically, the study examines the quality of growth by sectors, namely, the primary sector (growth rate in the agriculture sector); secondary sector (growth in the manufacturing sector); and the tertiary sector (growth in the services sector) to control for the bias introduced by using aggregate economic growth data (GDP) as seen in previous studies.

The study also investigated the role of institutional quality in the economic growth-poverty nexus since institutions are essential for durable growth (Olaoye and Aderajo 2020). The study adopts control of corruption as a measure of institutional quality. This is because the high level of corruption prevalent in sub-Saharan Africa has been identified as the single most important factor inhibiting development in the region (Gupta et al. 2009; Olaoye and Aderajo 2020).

Lastly, the study adopts a sample of similar developing countries (in this case, sub-Saharan Africa) to side-step the issues associated with pooled panel data (pooling developing, emerging, and industrialized economies into a single panel). The pooled panel data approach fails to address the region-specific and heterogeneity issues.

The results show that growth in the services sector (the knowledge-based sector) enhances green growth, growth and helps to reduce poverty in the region, while the agricultural and manufacturing sectors do not have significant poverty-reducing effects. This implies that for African countries to take the next leap and reduce the prevalence of poverty, the government must redirect the pattern of production and employment toward the services sector, which must be rapidly developed and expanded.

The rest of the paper is organized as follows: Sect. 3 describes the data. Section 4 contains the theoretical frame work, model specification and methodology. Section 8 presents the empirical findings, and Sect. 10 concludes the paper.

3 Data

The study used a balanced panel model for a sample of 40 sub-Saharan African countries (see Appendix Table 7). Secondary data spanning 1990–2019 are used in this study. Various proxies for poverty [extreme poverty (using $1.90)], $3.20 lower-middle income poverty line and a measure of multi-dimensional poverty) were adopted. Similarly, the study used the poverty gap index and the headcount ratio. However, the study emphasizes the poverty index gap, which measures the intensity of poverty. More precisely, unlike the headcount ratio, which considers the number of people living below the poverty line and considers them equally poor (Amartya 1976), the poverty gap index estimates the depth of poverty by looking at how far the poor are from the poverty line (Grusky and Ravi 2006).

Other variables are unemployment, institutional quality, government spending on health and education, population growth rate, financial development index, and income inequality. It is important to include income inequality since the level of inequality might dampen the impact of economic growth on poverty reduction (see Adeleye et al. 2020; Breunig and Majeed 2019; Fosu 2017). Trade openness, inflation, and interest rate are the control variables. See Appendix for the list of variables and data sources.

It is important to note that the data on poverty is scanty and not consistent. Thus, to side-step this problem, the study adopts a 6-year cumulative overlapping model following Bekaert et al. (2001, 2005), Cecchetti et al. (2011), and Checherita-Westpahl and Rother (2012). This approach helps to mitigate the issue of heterogeneity in growth regressions and also to control for business cycle phenomena. Besides, this approach is suitable for capturing the long-term impact of growth on poverty.

4 Conceptual Framework, Model Specification, and Methodology

No such theory directly connects model variables (growth and poverty reduction). Thus, the study presents the link between these two macroeconomic variables via the conceptual framework shown in Fig. 1.

The study adapts the conceptual framwork of Son and Kakwani (2004) which shows that alongside growth, policies that seek to redistribute income and assets equitably are important for poverty reduction. This implies a policy agenda that addresses both distributional concerns and economic growth could help to reduce poverty and promote equity.

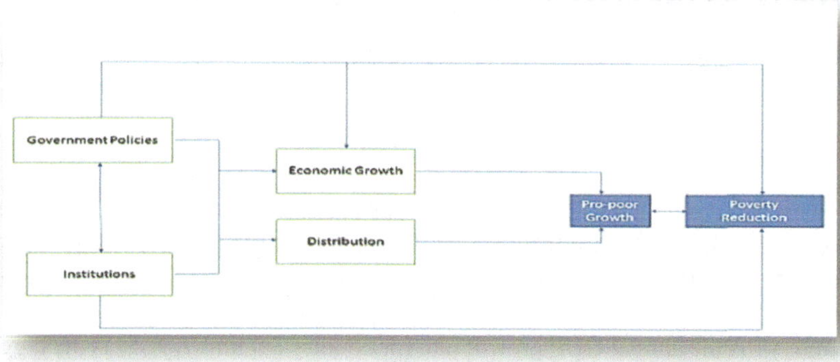

Fig. 1 Graphical illustration showing the link between economic growth and poverty reduction. *Source* Olaoye and Aderajo (2020)

5 Model Specification on the Effect of Economic Growth on Poverty Reduction

Following (Ravallion 1995, 1997; Easterly 2000; Rodrik 2000; Akbar 2004; Adeleye et al. 2020), the study specifies the model in a linear form as:

$$p_{it} = \tau p_{i,t-1} + \alpha_i + \omega y_{it} + X'_{it} + \eta C_{it} + \varepsilon_{it} \tag{1}$$

$$\varepsilon_{it} = \eta_t + \upsilon_{it}$$

Note: The error term contains both country/time-invariant effects and the idiosyncratic shocks. We assume that the error term is i.i.d with zero mean and constant variance.

Where subscript i is the country index and t is the time index, respectively, p is the log transformation[2] of poverty indices, X_{it} captures the explanatory variables, and these are population growth rate, institutional quality, unemployment rate, government spending on education and health; income inequality, and output gap. C is a vector of control variables which includes inflation and interest rate, α_i is a country-specific term, and ε_{it} is the usual error term. Equation (1) is the basic model of estimation. Through Eq. (1), the study investigates whether economic growth translates to poverty reduction.

[2] We log-transformed some of the variables to give the variables a constant variance and control for potential heteroskedasticity.

6 Model Specification on Drivers of Economic Growth

$$y_{it} = y_{i,t-1} + \alpha + X'_{it}\beta + \varepsilon_{it} \tag{2}$$

$$\varepsilon_{it} = \eta_t + \upsilon_{it}$$

For drivers of economic growth, the study follows extant studies (see ACBF 2017; Jaunky 2013; Mijiyawa 2013; IMF 2012; Ghazanchyan and Stotsky 2013) to include factors such as infrastructure, fiscal policy (captured by public debt), governance, and trade.

Where y_{it} is the annual GDP growth rate for country i at time t, $y_{i,t-1}$ is the lagged value of annual GDP growth rate and α is the model's constant term. X_{it} is a vector of independent variables (gross government debt/GDP, gross fixed capital formation/GDP, and human capital development measured by government spending on health and education. This is important because government spending on health and education has been described has important drivers of economic growth and poverty reduction. A higher level of education makes individuals more knowledgeable and increases their productivity. While investment in health improves life expectancy, increases productivity, enhances consumer earning power, and ultimately reduces poverty. See Appendix for the list of variables and data sources.

7 Estimation Techniques

7.1 Baseline Estimation Technique

The baseline estimation technique is the generalized method of moments (GMM) which accounts for simultaneity and endogeneity problems inherent in dynamic panel modeling. For more on the GMM estimator.

7.2 Cross-Sectional Dependence

For robustness, this study investigates whether the error terms in the panel models are cross-sectionally and spatially independent using Driscroll and Kraay (1998) robust standard errors. Previous studies have worked on the assumption that the error terms in panel data are serially uncorrelated. However, it is affirmed that panel data models are likely to show various cross-sectional and spatial dependencies (see Driscroll and Kraay 1998, Cameron and Trivedi 2005; Olaoye 2022b).

8 Preliminary Findings

The descriptive characteristics and stationarity properties of the variables are presented in Table 1. The correlation matrix presented in Table 1 shows that there is no issue of multicollinearity in the variables with correlation coefficients not higher than 0.5 except in the case of the different proxies of poverty which are intended to capture the same purpose.

The cross-sectional dependence test results are presented in Tables 2 and 3. The test is set under the null hypothesis (H_0); the errors are cross-sectional independent. The study applies a battery of cross-sectional dependence tests (see Table 2). Table 2 shows across all specifications that there is cross-sectional dependence in the model with *P*-Values <0.001. This suggests that previous estimates that ignored cross-sectional dependence might be unreliable, fallacious, and misleading.

Having established strong evidence of cross-section dependence, the study examined the stationary properties of the variables using second-generational panel unit root tests. The results are presented in Table 4. The results presented in Table 4 show that the variables are integrated at either level, i.e., $I(0)$, or at the first difference, i.e., $I(1)$. This suggests that the methodology adopted in this study is appropriate. Thus, in Table 8, in addition to the two-step system GMM, the study also adopts Driscoll-Kraay robust standard errors to correct for cross-sectional dependence and to account for the potential heterogeneity in economic growth-poverty reduction nexus (see Driscoll and Kraay 1998).

In estimating the Driscoll-Kraay robust standard errors, the study implements the pooled OLS (with robust standard errors). Interestingly, the Discroll-Kraay non-parametric covariance matrix estimator is adjusted for use with both balanced and unbalanced panels.

The study presents the results from the baseline model estimated via the system GMM (see Table 5). However, for reliable parameter estimates, the empirical analysis and conclusions are based on the pooled OLS (with robust standard errors) model estimates which accounts for cross-sectional dependency inherent in empirical panel data. Interestingly, the estimates from the pooled OLS appear to be more robust (see coefficient estimates in Table 5). The results show across all specifications that economic growth (captured by per capita GDP) exerts a positive or negative, albeit statistically insignificant, effect on poverty, indicating that economic growth recorded in sub-Saharan African countries under the period of study has not translated to a reduction in the level of poverty. This result might not be unconnected with the high level of corruption in the region and is consistent with the new institutional economics (NIE) theory.[3]

[3] The New Institution Economic (NIE) theory emphasizes the role of institutions as the fundamental determinant of economic and political development.

Table 1 Descriptive statistics of key variables

	$1.90	*$3.20*	*Multidi*	*CGD*	*Sp.health*	*Popula*	*Real.int*	*INF*	*UNEM*	*TAX*	*Sp.educ*	*CORR*	*TRADE*	*REER*	*pcapGDP*
Panel A															
Summary statistics															
mean	41.65	31.80	61.7	77.61	7.210	2.47	12.75	51.14	9.12	14.06	15.23	1.98	76.56	110.73	2.07
median	45.2	34.50	65.6	65.0	6.02	2.66	7.72	6.67	7.01	13.08	15.33	2.00	68.07	100.63	1.78
std. dev	0.18	15.9	7.06	67.09	4.43	1.09	63.43	409.1	7.75	6.6	7.66	0.77	35.09	45.75	4.96
observations	200	200	200	186	154	195	144	189	190	190	179	195	184	85	194
Panel B															
$1.90	1.000	0.962	0.945	0.205	0.321	0.490	0.358	0.400	−0.539	0.102	−0.390	−0.17	−0.470	0.324	0.045
$3.20		1.000	0.995	0.190	0.453	0.586	0.324	0.342	−0.681	0.521	−0.401	0.433	−0.619	0.241	−0.008
Multidi			1.000	0.170	0.124	0.564	0.307	0.323	−0.673	−0.78	−0.379	−0.61	−0.616	0.228	0.006
CGD				1.000	0.042	0.180	0.260	0.077	−0.231	−0.05	0.176	−0.36	−0.100	0.227	−0.334
Sp.health					1.000	0.232	0.126	0.089	0.154	0.123	0.31	−0.07	−0.298	0.262	0.045
Popula						1.000	0.257	0.060	−0.697	−0.42	−0.518	−0.70	−0.609	0.330	−0.145
Real.int							1.000	−0.132	−0.182	−0.05	−0.325	−0.36	−0.302	0.041	0.005
INF								1.000	0.002	0.124	−0.311	−0.15	−0.619	0.016	−0.220
UNEM									1.000	0.83	0.275	0.63	0.579	−0.189	0.139
TAX										1.000	0.3	0.080	0.462	−0.407	−0.243
Sp.educ											1.000	0.164	0.468	−0.045	−0.041
CORR												1.000	0.329	−0.115	0.064
TRADE													1.000	0.020	0.192
REER														1.000	− 0.048

(continued)

Table 1 (continued)

	$1.90	*$3.20*	*Multidi*	*CGD*	*Sp.health*	*Popula*	*Real.int*	*INF*	*UNEM*	*TAX*	*Sp.educ*	*CORR*	*TRADE*	*REER*	*pcapGDP*
pcapGDP															1.000

Source Authors' computations, 2022. *Notes* Table 1 shows the descriptive statistics and correlation matrix for 40 sub-Sahara African countries. $1.90, $3.20, and Multidi (Multidimensional) are measures of poverty using the poverty gap index. While INT, INF, CORR, CGD, Popula, Sp.health, UNEM, TAX, Sp.edu, TRADE. REER, pcapGDP, denote interest rate, inflation, control of corruption, central government debt (total debt), population (growth rate), government spending on healthcare (% of total expenditure), unemployment rate (ILO estimates), tax revenue, government spending on education (% of total expenditure), trade openness, real effective exchange rate, per capital GDP (growth rate), and output gap to capture economic shrinkage respectively. For the list of all the 40 SSA countries used in this study, see Appendix

Table 2 Panel cross-section dependence tests

Test	Results
1.Pesaran (2004)	21.876***(0.0000)
2.Breusch and Pagan (1979)	10.22***(0.000)
3.Pesaran (2007)	8.230***(0.0000)
4.Baltagi et al. (2012)	13.10***(0.0000)
5. Frees (2004)	2.832***(0.0000)
6. Friedman (1937)	157.805***(0.000)
7. **Pesaran (2004) CD**	14.410***(0.0000)
Average absolute value of the off-diagonal elements = 0.720	

Notes 1: Pesaran (2004) cross-sectional dependence in panel data models test; 2: Breusch-Pagan LM test of independence; 3: Baltagi et al. (2012) bias-corrected scaled LM test; 4: Pesaran (2007) bias-adjusted LM; 5: Frees (2004) test for cross-sectional dependence by using Frees' Q distribution (T-asymptotically distributed); 6: Friedman (1937) test for cross-sectional dependence using Friedman's χ^2 distributed statistics; 7: Pesaran (2004) CD test for cross-section dependence in panel time-series data; *P*-Values in parentheses. Tests include the intercept. ***$p < 0.01$, **$p < 0.05$, *$p < 0.10$

Table 3 Panel cross-section dependence tests (in the variable)

Variable	CD test	*P*-value	corr	abs (corr)
Poverty gap ($1.90)	15.37	0.0000	0.768	0.768
Poverty gap ($3.20)	17.32	0.0000	0.866	0.866
Per capital GDP growth	1.91	0.056	0.096	0.273
Corruption	14.42	0.000	0.868	0.868

Notes The cross-section dependence test is set under the null hypothesis of cross-section independence, CD ~ N(0,1) *P*-values close to zero indicate data are correlated across panel groups

Table 4 Second generational panel unit root tests

Panel unit root tests in presence of cross-section dependence			
	CIPS[a]	CADF[b]	
	Level		1st difference
Poverty gap ($1.90)	−5.769***	−4.393***	
Poverty gap ($3.20)	−5.670***	−4.226***	
Per capital GDP	−5.596***	−4.162***	
Corruption	–	−2.129	−3.009***
Unempl	−5.266***	−3.526***	
CGD	−3.034***	−3.082***	

Notes [a]H_0 (homogeneous non-stationary): $bi = 0$ for all I whereas; [b] the null hypothesis assumes all series are non-stationary in a heterogeneous panel with cross-sectional dependence. ***$p < 0.01$, **$p < 0.05$, *$p < 0.10$. Unempl and CGD denote Unemployment and Central Government Debt, respectively

Table 5 Dynamic panel estimates (system GMM)/Driscoll-Kraay non-parametric covariance matrix estimator adjusted for both balanced and unbalanced panel models

Non-overlapping	**Poverty gap $1.90** Cumulative 5-year models		**Poverty gap $3.20** Cumulative 5-year models		**Headcount ($1.90)** Cumulative 5-year models		**Headcount ($3.20)** Cumulative 5-year models		**Multi-dimensional Poverty** Cumulative 5-year models	
Variable	**Model 1** GMM	**Model 2** Pooled OLS (robust S.E.)	**Model 3** GMM	**Model 4** Pooled OLS (robust S.E.)	**Model 5** GMM	**Model 6** Pooled OLS (robust S.E.)	**Model 7** GMM	**Model 8** Pooled OLS (robust S.E.)	**Model 9** GMM	**Model 10** Pooled OLS (robust S.E.)
Poverty $1.90(−1)	0.8740*		–						–	
Poverty $3.20(−1)	–		0.2018						–	
Multi-dimensional(−1)	–		–		0.3714**		0.9246*			
pcapGDP	2.5165	0.0896*	5.4559	0.0692	4.6553	0.0025	−0.9132	−0.1151		2.6873
	(2.4417)	(0.0429)	(3.4853)	(0.0739)	(3.7453)	(0.1518)	(5.3421)	(0.1697)		(1.2789)
Sp. health	−0.0796	0.0066*	−0.0644	−0.0792	0.0031*	−0.0034	−0.0204	−0.0062*		−0.0161
	(1.431)	(0.0034)	(0.0428)	(1.034)	(0.0017)	(0.0866)	(0.2824)	(0.0031)		(0.2832)
Popula	3.6863	3.6760***	16.0564*	15.0564**	11.8879	6.9482*	10.5564	8.6496		3.5631
	(3.7298)	(1.024)	(5.6034)	(5.960)	(6.1627)	(3.2965)	(12.9445)	(4.6445)		(1.7593)
INF	0.0076*	−0.0003	0.0077**	0.0010	0.0093	−0.0032**	0.0185	0.0042**		0.1150*
	(0.0043)	(0.0011)	(0.0041)	(0.0013)	(0.0761)	(0.0010)	(0.1065)	(0.0013)		(0.0396)
UNEM	−0.3822	0.5778***	−1.861	−0.8904***	−2.3683	−0.9930**	0.1808	−1.0675***		−1.2015
	(0.6773)	(0.1364)	(0.7403)	(0.1791)	(0.9384)*	(0.2442)	(1.4846)	(0.2465)		(0.6807)
Sp. educ	−0.4189	−0.2789	0.6028	−0.4163*	1.7812	−0.4830	3.4548	−0.4479		−0.9136

(continued)

Table 5 (continued)

Non-overlapping	**Poverty gap $1.90** Cumulative 5-year models		**Poverty gap $3.20** Cumulative 5-year models		**Headcount ($1.90)** Cumulative 5-year models		**Headcount ($3.20)** Cumulative 5-year models		**Multi-dimensional Poverty** Cumulative 5-year models	
Variable	**Model 1** GMM	**Model 2** Pooled OLS (robust S.E.)	**Model 3** GMM	**Model 4** Pooled OLS (robust S.E.)	**Model 5** GMM	**Model 6** Pooled OLS (robust S.E.)	**Model 7** GMM	**Model 8** Pooled OLS (robust S.E.)	**Model 9** GMM	**Model 10** Pooled OLS (robust S.E.)
	(0.5525)	(0.1560)	(0.3083)	(0.1919)	(0.046)	(0.3049)	(2.0768)	(0.2662)		(0.5251)
Inequality	−0.0192*	0.2265***	0.0620***	0.1679***	−0.0701	−0.2265	−0.1597	−0.3045		0.1819***
	(0.0102)	(0.0421)	(0.0064)	(0.0312)	(0.0606)	(0.2409)	(0.2325)	(0.2432)		(0.0332)
Corr		−2.3625		−3.7141		−5.3774		−6.0944		−10.9766*
		(3.2666)		(4.6915)		(4.9201)		(5.2610)		(3.7163)
Inequality*pcapGDP		1.0028**		0.0822*						1.104*
Year and Country dummies (1990, 2008, 2010, 2014)	included									
Hansen	0.889		0.990		0.558					
AR(1)	0.052		0.011		0.041					
AR(2)	0.089		0.098		0.524					
R-squared		0.1524		0.2176		0.3031		0.3860		0.3566
Root MSE		10.5132		13.0140		17.1321		16.2403		19.2896

(continued)

Table 5 (continued)

Non-overlapping	**Poverty gap $1.90** Cumulative 5-year models		**Poverty gap $3.20** Cumulative 5-year models		**Headcount ($1.90)** Cumulative 5-year models		**Headcount ($3.20)** Cumulative 5-year models		**Multi-dimensional Poverty** Cumulative 5-year models	
Variable	**Model 1** GMM	**Model 2** Pooled OLS (robust S.E.)	**Model 3** GMM	**Model 4** Pooled OLS (robust S.E.)	**Model 5** GMM	**Model 6** Pooled OLS (robust S.E.)	**Model 7** GMM	**Model 8** Pooled OLS (robust S.E.)	**Model 9** GMM	**Model 10** Pooled OLS (robust S.E.)
F-Statistics		13.99		9.39		18.49		36.39		30.88
Prob (F-Statistics)		0.0117		0.0241		0.0069		0.0019		0.0317
Observations	195	139	195	139	160	140	195	140		44

Notes *, **,*** denote 10, 5 and 1% levels of significance, respectively. pcapGDP, Sp.health, Sp.educ., Popula, INF, UNEM, Corr. denote per capita GDP, government spending on health, government spending on education, population, inflation, unemployment and fight against corruption

9 Robustness Measure (Disaggregate Growth Model)

As a form of robustness and to control for the bias introduced by using aggregate economic growth data (GDP) as seen in previous studies, the study disaggregates economic growth into sectoral growths, namely, the primary sector (growth rate in the agriculture sector); secondary sector (growth in the manufacturing sector); and tertiary sector (growth in the service sector). The results are presented in Table 6. The results show that regardless of the measure of poverty, only the growth in the service sector shows a negative and significant impact on poverty (see column 4). This implies increasing the service sector[4] growth rate can significantly reduce the prevalence of poverty in the region. This implies for African countries to take the next leap, governments must review the patterns of employment and distribution toward the service sector, which must be rapidly developed and expanded.

10 Conclusions and Policy Recommendations

This paper examines the paradox of high economic growth and low poverty reduction rate for sub-Saharan African countries. Available evidence shows that many African countries have enjoyed strong economic growth over the last few decades. However, the benefits of economic growth have not trickle-down toward poor people. To this end, the study examined the quality of growth in SSA along the different sectors.

The result shows that an increase in the services sector growth rate enhances green and quality growth and leads to a decline in the level of poverty across the region. This implies the service sector must be rapidly developed and expanded if SSA is to take the next leap and reduce poverty prevalence.

Following the findings from the empirical study, this study makes the following recommendations:

One, for economic growth to translate to a reduction in the level of poverty across sub-Saharan African countries, must redistribute income through social protection programs (such as safety nets and social assistance transfers) to share the proceeds of growth equitably among the poor.

Two sub-Saharan African countries must redirect the pattern of production and employment toward a knowledge-based economy by developing the services sector (industries such as travel, Internet and information technology, media and entertainment, financial services, and healthcare). This would go a long way to reduce carbon emissions, create a safe environment, and ultimately reduce the level of poverty in SSA.

[4] By service sector, we refer to the economic sector that produces and offers services (such as travel, Internet and information technology, media and entertainment, financial services, and the healthcare industry). This sector is also referred to as the knowledge-based economy.

Table 6 A disaggregated growth model (Driscoll-Kraay non-parametric covariance matrix estimator adjusted for both balanced and unbalanced panel models) non-overlapping cumulative 5-year model; pooled OLS (robust S.E.)

	Poverty gap ($1.90)				*Poverty gap ($3.20)*				*Multi-dimensional poverty*			
Variable	***1***	**2**	3	***4***	5	***6***	7	***8***	***9***	***10***	***11***	***12***
growth (pcapGDP)	0.0896*				0.0629				2.6873			
	(0.0429)				(0.0739)				(1.2789)			
growth (aGRIC)		0.1718*				0.2058				0.6735*		
		(0.0740)				(0.1376)				(0.2381)		
growth (mANUFAC)			−0.0125				−0.0402				−0.0068	
			(0.0281)				(0.0313)				(0.1014)	
growth (sERVICE)				−0.0957*				−0.0954				−0.0069
				(0.0377)				(0.0937)				(0.2023)
Sp. health	0.0066*	–	–	–	−0.0792	–	–	–	−0.0161	–	–	–
	(0.0034)				(1.034)				(0.2832)			
Popula	3.6760***	3.4179**	3.4099**	4.1639**	15.0564**	6.5588	6.5131	6.4727	3.5631	5.9474	3.6442	3.6339
	(1.024)	(1.2956)	(1.3147)	(1.3193)	(5.960)	(2.5171)	(2.4762)	(2.4998)*	(1.7593)	(3.5924)	(3.7428)	(3.7948)
INF	−0.0003	−0.0010	−0.0011	−0.0013**	0.0010	−0.0021*	−0.0022**	−0.0023**	0.1150*	0.1658*	0.1427*	0.1424*
	(0.0011)	(0.0007)	(0.0007)	(0.0003)	(0.0013)	(0.0008)	(0.0007)	(0.0007)	(0.0396)	(0.0427)	(0.0416)	(0.0459)
UNEM	0.5778**	–	–	–	−0.8904*	–	–	–	−1.2015	–	–	–
	(0.1364)				(0.4791)				(0.6807)			
Sp. educ	−0.2789	−0.2172	−0.2031	−0.1260	−0.4163*	−0.1905	−0.1739	−0.1814	−0.9136	0.0013	−0.2710	−0.2738

(continued)

Table 6 (continued)

	Poverty gap ($1.90)				*Poverty gap ($3.20)*				*Multi-dimensional poverty*			
	(0.1560)	(0.1285)	(0.1590)	(0.1661)	(0.1919)	(0.1521)	(0.1835)	(0.2050)	(0.5251)	(0.6765)	(0.6144)	(0.6321)
Inequality	0.2265***	–	–	–	0.1679***	–	–	–	0.1819***	–	–	–
	(0.0421)				(0.0312)				(0.0332)			
Corr	−2.3625 (3.2666)	−2.0034 (2.3572)	−2.2912 (2.4669)	−2.5405 (2.0519)	−3.7141 (4.6915)	−3.8840 (2.8342)	−4.1302 (2.9582)	−4.1012 (2.9610)	−10.9766* (3.7163)	−8.3404 (4.4200)	−9.1791 (4.8207)	−9.1683 (5.1931)
R-squared	0.1524	0.2328	0.2154	0.1576	0.2176	0.2257	0.2113	0.2192	0.3566	0.3350	0.2751	0.2751
Root MSE	10.5132	10.0399	10.3388	10.4807	13.0140	12.9461	13.2858	13.0007	19.2896	19.6116	20.4751	20.4751
F-Statistics	13.99	30.02	15.64	7.63	9.39	4.77	2.96	5.25	30.88	34.68	25.52	41.92
Prob (F-Statistics)	0.0117	0.0027	0.0092	0.0347	0.0241	0.0761	0.1563	0.0651	0.0317	0.0283	0.0382	0.0235
Observations	139	139	134	139	139	139	134	139	44	44	44	44

*Notes**, **,*** denote 10, 5 and 1% levels of significance, respectively. growth (aGRIC), growth (mANUFAC), growth (sERVICE), Sp.health, Sp.educ., Popula, INF, UNEM, Corr. denote growth in agriculture sector, growth in manufacturing sector, growth in service sector, government spending on health, government spending on education, population, inflation, unemployment and fight against corruption

Lastly, governments in sub-Saharan Africa should strengthen economic and political institutions, including anti-corruption measures to reduce room for rent-seeking, combat corruption and improve domestic governance.

Appendix

See Table 7.

See Table 8.

Table 7 A list of 40 sub-Saharan African countries used in this study

Burkina Faso	Mozambique	Cameroon	Equatorial Guinea
Central African Republic	Niger	Congo	Gabon
Chad	Rwanda	Ghana	Namibia
Congo, DR	Senegal	Kenya	South Africa
Ethiopia	Tanzania	Lesotho	Mauritius
	Uganda	Nigeria	Seychelles
	Zimbabwe	Sao Tome and Principe	
	Benin	Swaziland	
	Eritrea	Zambia	
	Togo	Cape Verde	
	Sierra Leone	Cote d'Ivoire	
	Angola	Mauritania	
	Gambia, The	Botswana	
	Liberia		
	Malawi		
	Mali		

Oil exporters	Other commodity exporters	Non-commodity exporters
Angola	Botswana	Benin
Cameroon	Burkina Faso	Burundi
Chad	Central Africa Rep	Cabo Verde
Congo, Rep. of	Congo, Dem. Rep. of	Comoros
Equatorial Guinea	Ghana	Cote d'Ivoire
Gabon	Guinea	Eritrea
Nigeria	Liberia	Ethiopia

(continued)

Table 7 (continued)

Oil exporters	Other commodity exporters	Non-commodity exporters
South Sudan	Mali	Gambia, The
	Namibia	Guinea-Bissau
	Niger	Kenya
	Sierra Leone	Lesotho
	South Africa	Madagascar
	Tanzania	Malawi
	Zambia	Mauritius
	Zimbabwe	Mozambique
		Rwanda
		Sao Tome and Principe
		Senegal
		Seychelles
		Swaziland
		Togo
		Uganda

Table 8 Measurement and sources of key variables

Variables	Definition/Measurement	Source
1. Institutional quality captured by F-Corruption	This index measures a country's fight against corruption	ICRG
2. Real gross domestic product	It measures the productive capacity of the economy in reals terms	World Development Indicators database (WDI)
3. Capital stock	This is the general indicator for total capital stock	WDI
4. Trade Openness	It is a measure of the extent of trade openness and/or restriction	WDI
5. Poverty index	It measures the prevalence of poverty. We adopt different measures, such as the international poverty line of $1.90 for extreme poverty, the lower-and-middle-income class poverty line of $3.20, and the multi-dimensional poverty measure	WDI, World Bank, World Poverty Clock Data and Quartz Africa
6. Real interest rate	This is the bank lending rate adjusted for prices	WDI

(continued)

Table 8 (continued)

Variables	Definition/Measurement	Source
7. Inflation rate	This is measured by consumer prices	WDI
8. Unemployment	It measures the rate of unemployment in the country. ILO estimates	WDI
9. Income inequality	Captured by Gini coefficient. It measures how unevenly income is distributed among the population	WDI
10.Financial development	This measure captures financial depth, access, efficiency and stability	IMF, International Financial Statistics, Financial Development Index Database
11. Output gap	It is used to measure economic shrinking or business fluctuations. It is measured using the Hodrick-Prescott technique for separating the short from the long term. We use the Hodrick-Prescott filtering technique to filter the GDP series (with a smoothing parameter of 6.25 as recommended by Ravn and Uhlig (2002) for annual figures)	Author's computation
12.Government spending on health and education	It measures the total amount allocated to education and health as a percentage of total government spending	WDI, National estimates

References

ACBF (2017) Drivers of economic growth in Africa, Occasional Paper No. 29, African Capacity Building Foundation, Harare, Zimbabwe

Adejumo OO, Adejumo AV, Aladesanmi TA (2020) Technology-driven growth and inclusive growth-implications for sustainable development in Africa. Technology in Society 63. Available at https://doi.org/10.1016/j.techsoc.2020.101373

Adeleye BN, Gershon O, Ogundipe A, Owolabi O, Ogunrinola I, Adediran O (2020) Comparative investigation of the growth-poverty-inequality trilemma in Sub-Saharan Africa and Latin American and Caribbean Countries. Heliyon 6(2020):e05631

Agrawal P (2008) Economic growth and poverty reduction: evidence from Kazakhstan. Asian Dev Rev 24(2):90–115

Akbar M (2004) Is poverty rising in Pakistan. DAWN—business, October 11, DAWN the internet edition [online; accessed Nov 11, 2011]. Available from http://archives.dawn.com/2004/10/11/ebr3.htm

Amartya S (1976) Poverty: an ordinal approach to measurement. Econometrica 44(2):219–231. https://doi.org/10.2307/1912718

Arndt C, McKay A, Tarp F (eds) (2016) Growth and poverty in sub-saharan africa. Oxford: Oxford University Press
Bhagwati J (2001) Trade and poverty in the poor countries. Am Econ Rev, Am Econ Assoc 92(2):180–183
Baloch MA, Khan SUD, Ulucak ZS, Ahmad A (2020) Analyzing the relationship between poverty, income inequality, and CO_2 emission in Sub-Saharan African countries. Available at https://doi.org/10.1016/j.scitotenv.2020.139867
Bekaert G, Harvey CR, Lundblad C (2001) Emerging equity markets and economic development. J Dev Econ 66(2):465–504
Bekaert G, Harvey CR, Lundblad C (2005) Does financial liberalization spur growth? J Financ Econ 77(1):3–55
Breunig R, Majeed O (2020) Inequality, poverty and economic growth. Int Econ 161(C):83–99
Cameron T, Trivedi PK (2005) Microeconometrics: methods and applications. Cambridge: Cambridge university press. https://doi.org/10.1017/CBO9780511811241
Cecchetti SG, Mohanty MS, Zampolli F (2011) The real effects of debt, BIS working papers no. 352, Sept 2011
Checherita-Westpahl C, Rother P (2012) The impact of high government debt on economic growth and its channels: an empirical investigation for the euro area. Eur Econ Rev 56(7):1392–1405
Danquah M, Iddrisu A M, Quartey P, Ohemeng W, Barimah A (2022) Rural financial intermediation and poverty reduction in Ghana: a micro-level analysis. Poverty Public Policy 13(4):316–334
Datt G, Ravillion M (2002) Is India's economic growth leaving the poor behind? Journal of Economic Perspectives 16(3):89–108
Deaton A, Dreze J (2001) Trade, growth and poverty. Finance and Development 38:16–19
Dollar D, Kraay A (2000) Growth is good for the poor (preliminary and incomplete). World Bank, Washington
Dollar D, Kraay A (2001) Growth is good for the poor. J Econ Growth 7(1):195–225
Dollar D, Kraay A (2002) Growth is good for the poor. J Econ Growth 7(1):195–225
Donaldson JA (2008) Growth is good for whom, when, how? Economic growth and poverty reduction in exceptional cases. World Dev 36(11):2127–2143
Driscoll JC, Kraay AC (1998) Consistent covariance matrix estimation with spatially dependent panel data. Rev Econ Stat 80:549–560
Easterly W (2000) The effect of IMF and World Bank programs on poverty. World Bank, Washington
Erlando A, Riyanto FD, Masakazu S (2020) Financial inclusion, economic growth, and poverty alleviation: evidence from eastern Indonesia. Heliyon 6(10). https://doi/10.1016/j.heliyon.2020.e05235
Ferreira FHG, Leite PG, Ravallion M (2010) Poverty reduction without economic growth? Explaining Brazil's poverty dynamics, 1985–2004. J Dev Econ 93:20–36
Fosu AK (2017) Growth, inequality, and poverty reduction in developing countries: recent global evidence. Res Econ 71:306–336
Fosu AK, Gafa DW (2020) Progress on poverty in Africa: how have growth and inequality mattered? Afr Rev Econ Finan 12(1)
Ghazanchyan M, Stotsky J (2013) Drivers of growth: evidence from Sub-Saharan African countries. IMF working paper WP/13/236, International Monetary Fund, Washington, DC
Grusky DB, Ravi K (Ed.) (2006) Poverty and inequality. Stanford University Press
Gupta S, Pattillo CA, Smita W (2009) Effect of remittances on poverty and financial development in sub-saharan africa. World Dev 37(1):104–115
Hailu KK, Nagaraja G (2017) Impact of growth and inequality on poverty in Ethiopia: empirical evidence from Amhara region. Int J Appl Res 3(7):78–683
IMF (2012) Maintaining growth in an uncertain World, Chap. 1. In: Regional economic outlook, Sub-Saharan Africa, October, International Monetary Fund, Washington, DC, pp 1–23
International Debt Statistics (2020) World Bank Group. Washington, DC. http://www.worldbank.org

Jaunky V (2013) Democracy and economic growth in Sub-Saharan Africa: a panel data approach. Empirical Economics 45(2):987–1008

Khan SAR (2019) The nexus between carbon emissions, poverty, economic growth, and logistics operations-empirical evidence from southeast Asian countries. Environmental Science and Pollution Research. Available at https://doi.org/10.1007/s11356-019-04829-4

Kudebayeva A, Barrientos A (2013) A decade of poverty reduction in Kazakhstan 2000–2009: growth and/or redistribution? Available at http://ssrn.com/abstract=2334179

Mijiyawa A (2013) Africa's recent economic growth: what are the contributing factors? Afr Dev Rev 25(3):289–302

Moore JD, Donaldson JA (2016) Human-scale economics: economic growth and poverty reduction in northeastern Thailand. World Dev. https://doi.org/10.1016/j.worlddev.2016.04.004

Mulok D, Kogid M, Asid R, Lily J (2012) Is economic growth sufficient for poverty alleviation? Empirical evidence from Malaysia. Cuadernos De Economía 35:26–32

Olaoye OO, Aderajo OM (2020) Institutional Infrastructure and Economic Growth in ECOWAS: an investigation into the hierarchy of institution hypothesis (HIH). Int J Soc Econ 47(9):1081–1108

Olaoye OO (2022a) Sub-Saharan Africa's debt-financed growth: how sustainable and inclusive? Afr Dev Rev. https://doi.org/10.1111/1467-8268.12670

Olaoye, O. O. (2022b) Analysis of the prevalence of extreme poverty in Africa south of the Sahara: do geographical proximity and institutional spillovers matter? Journal of Economic Studies. https://doi.org/10.1108/JES-07-2021-0322

Ravallion M (1995) Growth and poverty: evidence for developing countries in the 1980s. Econ Lett 48(3–4):411–417

Ravallion M (1997) Can high inequality developing countries escape absolute poverty? Econ Lett 56:1–57

Ravallion M (2002) Why don't we see poverty convergence? American Economic Review 102(1):504–523

Ravn M, Uhlig H (2002) On Adjusting the Hodrick-Prescott filter for the frequency of observations. Rev Econ Stat 84(2):371–376

Rodrik D (2000) Growth versus poverty reduction: a hollow debate. Finance and Development 37(4):8–9

Son HH, Kakwani NC (2004) Economic growth and poverty reduction: initial conditions matter. United nations development programme, Working Paper number 2

The World Bank (2015) Climate change complicates efforts to end poverty. Scribbr. https://www.worldbank.org/en/news/feature/2015/02/06/climate-change-complicates-efforts-end-poverty

Financial Inclusion and Environmental Sustainability

Peterson K. Ozili

1 Introduction

In this paper, I examine the association between financial inclusion and environmental sustainability. I predict that financial inclusion programs and efforts can complement certain environmental sustainability efforts toward achieving the United Nations sustainable development goals (SDGs). This is because financial inclusion strategies can be designed in ways that ensure that formal finance is accessible to all members of society while at the same time, ensuring that users of formal finance use formal financial services in ways that promote the continuity of society and the preservation of the environment. To test the validity of this argument, I empirically examine whether there is a direct link or association between financial inclusion and environmental sustainability using several indicators of financial inclusion and environmental sustainability.

The previous research on financial inclusion and sustainability suggests that financial inclusion can contribute to sustainable development. For example, Le et al. (2019a, b) examine the impact of financial inclusion on financial efficiency and financial sustainability in Asia. They find that high levels of financial inclusion have a positive impact on financial sustainability. Arner et al. (2020) show that the presence of infrastructure to support the digital financial transformation of Fintech is vital to achieve the SDGs. Machdar (2020) analyzes the effect of financial inclusion on sustainable economic growth and shows that financial inclusion does not affect sustainable economic growth in banks in Indonesia. These studies did not examine the association or correlation between financial inclusion efforts and environmental sustainability efforts toward achieving the sustainable development goals.

Using data for 27 OECD countries from 2004 to 2012, and focusing on some financial inclusion and environmental sustainability indicators, the main findings suggest

P. K. Ozili (✉)
Governors Department, Central Bank of Nigeria, Abuja, Nigeria
e-mail: petersonkitakogelu@yahoo.com

M. El Amine Abdelli and M. Shahbaz (eds.), *Digital Economy, Energy and Sustainability*, Green Energy and Technology, https://doi.org/10.1007/978-3-031-22382-2_2

that financial inclusion is positively associated with environmental sustainability, and the association is stronger among non-EU countries.

This study contributes to the literature in the following ways. One, it contributes to the literature on sustainable development (Kartick 2012; Arner et al. 2020; Ozili 2020b, 2021b). It contributes to this literature by showing how financial inclusion efforts and environmental sustainability efforts complement each other. Two, this study contributes to the financial inclusion literature (Datta and Singh 2019; Ozili 2018, 2020a, 2021a; De Matteis 2015; Didenko et al. 2020). It contributes to this literature by showing that financial access indicators, such as the availability of ATM facility and bank branch networks, have positive effects for sustainable development. Three, this study contributes to the literature on environmental sustainability. It contributes to this literature by establishing a link between environmental policies and financial development.

The rest of the paper is structured as follows. Section 2 presents the conceptual framework. Section 3 presents the literature review. Section 4 presents the research design and the descriptive statistics. Section 5 discusses the results. Section 6 concludes.

2 Conceptual Framework

Financial inclusion and environmental sustainability are two concepts that have been widely linked to sustainable development (Yin et al. 2019; Klapper et al. 2016). Financial inclusion and environmental sustainability can help to achieve the sustainable development goals (SDGs) from two different pathways. Specifically, the financial inclusion agenda contributes to sustainable development by reducing global poverty levels through the provision of affordable financial services to the excluded population (Chibba 2009; Park and Mercado 2018), while the environmental sustainability agenda contributes to sustainable development by reducing climate change through environmental protection policies (Haque and Ntim 2018; Yuan and Zhang 2020). Strict environmental policies and environment-adjusted growth can ensure that economic growth is achieved without sacrificing the environment. This suggest that there is some complementarity between financial inclusion and environmental sustainability as they both contribute to achieving the SGDs even though they do so from two different dimensions—one from a financial services dimension and the other from an environmental protection dimension. Figure 1 illustrates the complementarity between the two concepts.

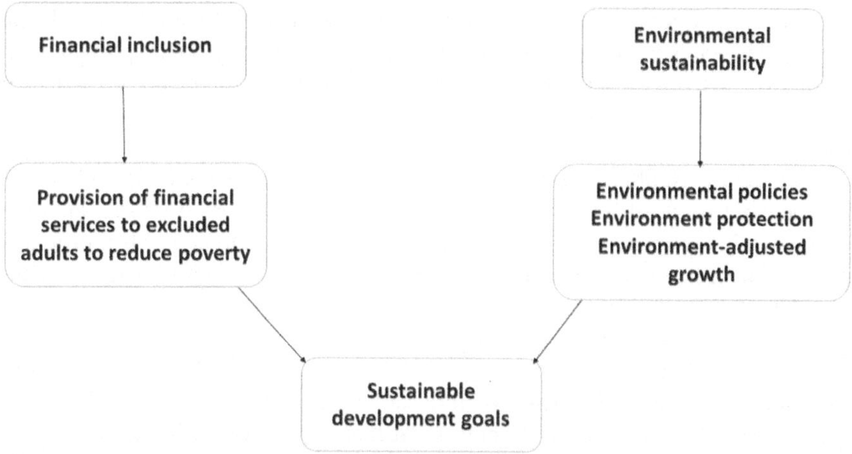

Fig. 1 Linking financial inclusion and environmental sustainability to sustainable development. *Source* Author

3 Literature Review

3.1 Financial Inclusion

Voica (2017) argues that financial inclusion is an important tool to promote sustainable development because innovations in financial services such as microfinance can help to uplift people out of poverty. Kartick (2012) argues that inclusive growth cannot be achieved without high levels of financial inclusion. Yin et al. (2019) show that economic policies, such as monetary policy, have a short-term positive impact on financial inclusion factors, which implies that expansionary monetary policy can improve the level of financial inclusion in the short term.

In the United Kingdom, Rahim et al. (2009) find that the vast majority of new migrants in the United Kingdom have low incomes and are more vulnerable to financial exclusion because they face difficulties in accessing and using financial services. These difficulties are caused by both their legal status, as well as supply-side and demand-side factors. They stress that, to remove these barriers, banks should engage new migrants by using outreach programs and tailoring their financial services to meet the needs of new migrants. Also, banks should produce multi-language marketing materials, recruit multilingual staff, and reduce the unnecessary identification requirements. Also, Lewis and Lindley (2015) show that new forms of financial exclusion are emerging as digital technology advances, and risk profiling becomes increasingly sophisticated in the United Kingdom. Self-employed people face high levels of unsecured debt and are less likely to have pension savings. Financial capability is low among the United Kingdom population, and most often, the groups with the lowest financial capability are also at most risk of financial exclusion. They recommend that policy makers should provide coordination for financial inclusion policies, provide support for teaching financial education in schools, provide more progressive savings incentives, encourage banks to offer basic banking services to meet the needs of the most vulnerable, and streamline government support for small businesses.

In Germany, Neuberger (2015) shows that the quality of financial inclusion is impaired by a low level of financial literacy, and most vulnerable individuals and small enterprises in Germany tend to be excluded or credit constrained. Also, programs involving microfinance institutions have been stopped. They argue that authorities should implement the right to a basic bank account with an overdraft facility, establish public credit bureaus, redirect banking regulation toward the protection of borrowers in long-term credit relationships, strengthen financial education and entrepreneurship education in schools, and develop venture capital markets.

In the EU, De Matteis (2015) points out that migrants residing in the EU are deeply affected by economic crisis, and they are exposed to social and financial exclusion. De Matteis shows that community development schemes such as the 'self-funded communities model' can increase the level of financial inclusion for migrants residing in the EU.

In terms of financial inclusion programs, the number of bank branch per 100,000 adults and the number of ATMs per 100,000 adults are widely considered to be measures of supply-side financial accessibility, leading to greater financial inclusion. For example, the financial inclusion index developed by Lenka and Bairwa (2016) was derived from a combination of several variables including bank branch per 100,000 adults and ATMs per 100,000 adults. In their analysis, they find that financial inclusion is negatively related to the inflation rate, which suggest that high levels of financial inclusion lowers the level of inflation. Also, Didenko et al. (2020) show that an increase in ATMs per 100,000 adults lead to a significant increase in the social safety index, which suggest that people feel safer when they have better access to ATM facilities to withdraw money whenever they need money.

3.2 *Environmental Sustainability*

Johnstone et al. (2012) investigate the impact of public environmental policy on innovations in environment-related technology. They examine 77 countries between 2001 and 2007, and they find that environmental policy has a positive impact on environment-related innovation. Similarly, Morales-Lage et al. (2016) investigate the impact of environmental policy stringency on innovation and productivity in 14 OECD countries. They find that more stringent environmental regulations have a positive influence on research and development (R&D) expenditure, the number of patent applications, and total factor productivity.

Albrizio et al. (2014) show that the tightening of environmental policies has a positive effect on productivity growth at the firm level in technologically advanced

Table 1 Variable description and source

Variable	Description	Source
EP	Environmental policy stringency index	OECD statistics
EG	Environmentally adjusted multifactor productivity growth	Green growth indicators, OECD Statistics
CD	ATMs per 100,000 adults	Global financial development indicators, World Bank database
BR	Bank branches per 100,000 adults	Global financial development indicators, World Bank database

OECD countries. Meanwhile, Ahmed and Ahmed (2018) examine the effect of environmental policy stringency on economic activity in China. They find that stringent environmental policies reduce CO_2 emissions, and stringent environmental policies have a negative impact on GDP in China. This implies that strict environmental policies have a positive effect on environmental sustainability through a reduction in CO_2 emissions even though it has a negative effect on economic growth.

4 Research Design

4.1 Data

Financial inclusion data were collected from the global financial development indicators. Environmental sustainability data were collected from the OECD statistics database. Data were extracted for 27 OECD countries from 2004 to 2012. The reason for the narrow sample period is because financial inclusion data became available only from 2004 onwards; meanwhile, sustainability data were reported much earlier. See Table 1 for variables description and source.

4.2 Method of Analysis

The statistical technique used to measure the association between the financial inclusion indicators and the selected environmental sustainability indicators is the Pearson correlation test statistic. The Pearson correlation test statistic is widely used to measure the strength of the association between two variables (Gujarati and Porter 1999; Piaw 2013). It shows the strength of the linear correlation or association between two variables and their association with each other. The Pearson coefficient yields a positive or negative association between two variables, and its values range between -1 and $+1$. (Gujarati and Porter 1999).

4.3 Variable Justification

Financial inclusion was measured using two indicators: (i) bank branch per 100,000 adults (BR) and (ii) ATM per 100,000 adults (CD). The bank branch per 100,000 adults variable (BR) is commonly used in the literature to measure the level of financial inclusion (see, Neaime and Gaysset 2018; Naumenkova et al. 2019; Raza et al. 2019). A high number of bank branch per 100,000 adults in a country signifies high levels of financial inclusion in the country (Neaime and Gaysset 2018; Taddese Bekele and Abebaw Degu 2021). The ATM per 100,000 adults variable (CD) is also a widely used indicator of financial inclusion in the literature (Van and Linh 2019; Van et al. 2021). A high number of ATM per 100,000 adults in a country signifies high a level of financial inclusion in the country, and this is supported by Datta and Singh (2019), Le et al. (2019a, b), Van and Linh (2019), and Van et al. (2021).

Environmental sustainability is measured using two indicators: (i) the environmental policy stringency index (EP) and (ii) the environmentally adjusted multifactor productivity growth (EG). The environmental policy stringency index (EP) is a widely used indicator of environmental sustainability in the literature (see Galeotti et al. 2020; Martínez-Zarzoso et al. 2019). It measures the degree to which environmental policies put an explicit or implicit price on pollution or environmentally harmful behavior. A high environmental policy stringency index in a country indicates the strictness of environmental rules toward preserving the environment for sustainable development for today and future generations. The environmentally adjusted multifactor productivity growth (EG) index is a widely used indicator of environmental sustainability in the literature (see Gu et al. 2019; Tolliver et al. 2021). EG is used as a proxy for sustainability because it is a green growth headline indicator, and because it measures a country's ability to generate income from a given set of inputs while accounting for the consumption of natural resources and production of undesirable environmental outputs (Rodríguez et al. 2018). A high EG index in a country indicates higher green growth toward environmental sustainability.

4.4 Descriptive Statistics

The descriptive statistic is reported in Table 2. The environmental policy stringency index (EP) on average is 2.57, and EP is highest in Denmark and lowest in Turkey. Denmark records the highest average EP and is a EU country, while Turkey has the lowest EP average and is a non-EU country. This suggests that environmental policies are stronger within the EU than outside the EU. Also, the environmentally adjusted multifactor productivity growth (EG) is 1.26 on average, and EG is highest in Korea and lowest in Greece. Korea records the highest average EG and is a non-EU country, while Greece has the lowest EG average and is a EU country. This suggests that green growth is higher in countries outside the EU such as in Korea. ATMs per 100,000 adults (CD) on average is 99.44, and CD is highest in Korea and lowest in

Czech Republic. This suggests that financial inclusion (through ATMs penetration) is higher in countries outside the EU. Bank branch per 100,000 adults (BR) is 33.52 on average, and BR is highest in Spain and lowest in Norway. This suggests that bank branch penetration is high in the EU.

5 Results

5.1 *Pearson Correlation Results: Full Sample*

The full sample Pearson correlation result is reported in Table 3. Only the significant correlation coefficient results are interpreted. The two sustainability indicators (EP and EG) are significant and negatively correlated at −0.18***. This indicates that high environmental policy stringency is associated with lower environmentally adjusted multifactor productivity growth. This implies that strict environmental protection policies may harm green growth.

Also, the two financial inclusion indicators (BR and CD) are significant and positively correlated at 0.24***. This suggests that a higher number of ATM per 100,000 adults is associated with a larger number of bank branches per 100,000 adults, which jointly have positive effects for financial inclusion.

Interestingly, the EP and CD correlation coefficient (0.158***) is significant and positively correlated. This indicates that higher environmental policy stringency is associated with greater financial inclusion through higher number of ATMs per 100,000 adults. This implies that environmental sustainability is positively associated (or correlated) with financial inclusion.

On the other hand, the CD and EG coefficient (0.002), BR and EG coefficient (−0.091), and BR and EP coefficient (−0.013) are all statistically insignificant. Therefore, no meaningful conclusion can be drawn for this category.

5.2 *Subsample Analysis*

5.2.1 Correlation: European Countries Only

The Europe category correlation result is reported in Table 4. Only the significant correlation coefficient results are interpreted. The two environmental sustainability indicators (EP and EG) are also significant and negatively correlated at −0.22***, which indicates that high environmental policy stringency is associated with lower environmentally adjusted multifactor productivity growth in European countries.

Also, the two financial inclusion indicators (BR and CD) are also significant and positively correlated at 0.61***. This suggests that a higher number of ATM per 100,000 adults is associated with a larger number of bank branches per 100,000

Table 2 Descriptive statistic

S/N	Countries	EP	EG	CD	BR	Region
1	Australia	2.36	1.17	156.86	31.01	Non-EU
2	Austria	2.90	1.45	111.63	13.32	EU, Europe
3	Belgium	2.39	1.13	87.95	49.42	EU, Europe
4	Canada	2.91	1.04	207.21	24.36	Non-EU, G7
5	Czech Republic	2.48	1.83	39.01	22.34	EU, Europe
6	Denmark	3.40	0.81	65.80	45.84	EU, Europe
7	Finland	3.04	1.04	53.20	15.09	EU, Europe
8	France	3.11	1.15	100.02	38.06	EU, Europe, G7
9	Germany	2.91	1.43	114.25	16.93	EU, Europe, G7
10	Greece	2.00	−1.31	75.20	39.04	EU, Europe
11	Hungary	2.57	0.87	50.68	16.33	EU, Europe
12	Ireland	2.02	4.89	90.37	31.51	EU, Europe
13	Italy	2.5	0.57	91.51	58.33	EU, Europe, G7
14	Japan	2.08	1.08	127.01	34.03	Non-EU, G7
15	Korea	3.07	2.37	244.09	17.98	Non-EU
16	Netherlands	3.14	0.83	59.45	26.08	EU, Europe
17	Norway	2.51	1.11	57.43	11.54	EU, Europe
18	Poland	2.38	1.80	41.43	30.55	EU, Europe
19	Portugal	2.38	0.51	176.31	65.29	EU, Europe
20	Slovakia	2.03	1.98	47.82	26.25	EU, Europe
21	Slovenia	2.12	1.67	95.64	39.51	EU, Europe
22	Spain	2.76	0.87	149.32	97.02	EU, Europe
23	Sweden	2.98	1.07	40.94	23.04	EU, Europe
24	Switzerland	2.67	1.37	93.80	53.61	Non-EU, Europe
25	Turkey	1.53	0.89	43.71	16.19	Non-EU, Europe
26	United Kingdom	2.61	1.16	122.82	25.83	EU, Europe, G7
27	United States	2.26	1.31	169.67	34.56	Non-EU, G7
	Full sample:					
	Mean	2.57	1.26	99.44	33.52	
	Median	2.60	1.45	90.29	28.65	
	Maximum	4.13	14.25	288.47	104.21	
	Minimum	0.83	−7.01	25.44	9.92	
	Std. Dev	0.65	2.34	54.81	19.02	
	Observation	239	242	235	241	

Table 3 Pearson correlation statistic (all countries)

Variable	EP	EG	CD	BR
EP	1.000			
	–			
	–			
EG	−0.185***	1.000		
	(−2.83)	–		
	((0.005))	–		
CD	0.158**	0.002	1.000	
	(2.41)	(0.02)	–	
	((0.016))	((0.980))	–	
BR	−0.013	−0.091	0.241***	1.000
	(−0.19)	(−1.38)	(3.74)	–
	((0.842))	((0.167))	((0.000))	–

EP - Environmental policy stringency index. *EG* - Environmentally adjusted multifactor productivity growth index. *CD* - ATMs per 100,000 adults. *BR* - Bank branches per 100,000 adults. T-statistic is reported in single parenthesis. *P*-value is reported in double parenthesis. *** and ** denote statistical significance at the 1 and 5% levels. The countries in the full sample category are: Australia, Austria, Belgium, Canada, Czech Republic, Denmark, Finland, France, Germany, Greece, Hungary, Ireland, Italy, Japan, Korea, Netherlands, Norway, Poland, Portugal, Slovakia, Slovenia, Spain, Sweden, Switzerland, Turkey, United Kingdom, and the United States

Table 4 Pearson correlation analysis (Europe category)

Variable	EP	EG	CD	BR
EP	1.000			
	–			
	–			
EG	−0.221***	1.000		
	(−3.10)	–		
	((0.002))	–		
CD	0.107	−0.057	1.000	
	(1.47)	(−0.78)	–	
	((0.142))	((0.431))	–	
BR	0.029	−0.081	0.613***	1.000
	(0.39)	(−1.10)	(10.58)	–
	((0.693))	((0.270))	((0.000))	–

EP - Environmental policy stringency index. *EG* - Environmentally adjusted multifactor productivity growth index. *CD* - ATMs per 100,000 adults. *BR* - Bank branches per 100,000 adults. T-statistic is reported in single parenthesis. *P*-value is reported in double parenthesis. *** denotes statistical significance at the 1% level. There are 22 countries in the Europe category, namely Austria, Belgium, Czech Republic, Denmark, Finland, France, Germany, Greece, Hungary, Ireland, Italy, Netherlands, Norway, Poland, Portugal, Slovakia, Slovenia, Spain, Sweden, Switzerland, Turkey, and the United Kingdom adults, which jointly have positive effects for financial inclusion in European countries.

On the other hand, the CD and EP coefficient (0.107), CD and EG coefficient (−0.057), BR and EG coefficient (−0.081), and the BR and EP coefficient (0.029) are all statistically insignificant. Therefore, no meaningful conclusion can be drawn for this category.

5.2.2 Correlation: EU Countries Only

Not all countries in Europe are member-countries of the European Union (EU). In this section, I analyze the correlation for the EU member-counties in the OECD sample.

The EU countries correlation result is reported in Table 5. Only the significant correlation coefficient results are interpreted. The two sustainability indicators (EP and EG) are significant and negatively correlated at −0.255***. This indicates that high environmental policy stringency is associated with lower environmentally adjusted multifactor productivity growth in EU member-countries.

Also, the two financial inclusion indicators (BR and CD) are significant and positively correlated at 0.596***. This suggests that a higher number of ATM per 100,000 adults is associated with higher bank branches per 100,000 adults, which jointly have positive effects for financial inclusion in EU member-countries.

Table 5 Pearson correlation analysis (EU category)

Variable	EP	EG	CD	BR
EP	1.000			
	–			
	–			
EG	−0.255***	1.000		
	(−3.42)	–		
	((0.001))	–		
CD	0.009	-0.065	1.000	
	(0.12)	(-0.84)	–	
	((0.901))	((0.402))	–	
BR	−0.054	-0.097	0.596***	1.000
	(−0.69)	(−1.27)	(9.63)	–
	((0.485))	((0.204))	((0.000))	–

EP - Environmental policy stringency index. *EG* - Environmentally adjusted multifactor productivity growth. *CD* - ATMs per 100,000 adults. *BR* - Bank branches per 100,000 adults. T-statistics are reported in single parenthesis. *P*-values are reported in double parenthesis. *** denotes statistical significance at the 1% level. There are 20 countries in the EU category, namely Austria, Belgium, Czech Republic, Denmark, Finland, France, Germany, Greece, Hungary, Ireland, Italy, Netherlands, Norway, Poland, Portugal, Slovakia, Slovenia, Spain, Sweden, and United Kingdom

On the other hand, the CD and EP coefficient (0.009), CD and EG coefficient (−0.065), BR and EG coefficient (−0.097), and the BR and EP coefficient (−0.054) are all statistically insignificant. Therefore, no meaningful conclusion can be drawn for this category.

5.2.3 Correlation: Non-EU Countries Only

The non-EU country correlation result is reported in Table 6. Only the significant correlation coefficient results are interpreted. Also, the two financial inclusion indicators (BR and CD) are weakly significant and negatively correlated at 0.243***. This suggests that a higher number of ATM per 100,000 adults is associated with a fewer number of bank branches per 100,000 adults in non-EU countries.

Interestingly, the EP and CD correlation coefficient (0.585***) is significant and positively correlated. This indicates that higher environmental policy stringency is associated with greater financial inclusion through higher number of ATMs per 100,000 adults. This implies that sustainability is positively associated (or correlated) with financial inclusion for countries outside the European Union.

On the other hand, the CD and EG coefficient (0.096), BR and EG coefficient (−0.038), and the BR and EP coefficient (0.058) are all statistically insignificant. Therefore, no meaningful conclusion can be drawn for this category.

Table 6 Pearson correlation analysis (non-EU category)

Variable	EP	EG	CD	BR
EP	1.000			
	–			
	–			
EG	−0.016	1.000		
	(−0.12)	–		
	((0.899))	–		
CD	0.585***	0.096	1.000	
	(5.41)	(0.72)	–	
	((0.000))	((0.471))	–	
BR	0.058	−0.038	−0.243*	1.000
	(0.43)	(−0.29)	(−1.88)	–
	((0.663))	((0.772))	((0.065))	–

EP - Environmental policy stringency index. *EG* - Environmentally adjusted multifactor productivity growth. *CD* - ATMs per 100,000 adults. *BR* - Bank branches per 100,000 adults. T-statistics are reported in single parenthesis. *P*-values are reported in double parenthesis. *** and * denote statistical significance at the 1 and 10% levels. There are seven (7) countries in the non-EU category, namely Australia, Canada, Japan, Korea, Switzerland, Turkey, and the United States

Table 7 Pearson correlation analysis (G7 category)

Variable	EP	EG	CD	BR
EP	1.000			
	–			
	–			
EG	−0.178	1.000		
	(−1.31)	–		
	((0.195))	–		
CD	0.146	−0.033	1.000	
	(1.07)	(−0.24)	–	
	((0.289))	((0.809))	–	
BR	−0.139	−0.105	−0.441***	1.000
	(−1.02)	(−0.76)	(−3.54)	–
	((0.313))	((0.449))	((0.001))	–

EP - Environmental policy stringency index. *EG* - Environmentally adjusted multifactor productivity growth. *CD* - ATMs per 100,000 adults. *BR* - Bank branches per 100,000 adults. T-statistic is reported in single parenthesis. *P*-value is reported in double parenthesis. *** denotes statistical significance at the 1% levels. There are seven (7) countries in the G7 category, namely Canada, Japan, Italy, Germany, France, United Kingdom, and the United States

5.2.4 Correlation: G7 Countries Category

The G7 countries correlation result is reported in Table 7. Only the significant correlation coefficient results are interpreted. The two financial inclusion indicators (BR and CD) are significant and negatively correlated at −0.441***. This suggests that a higher number of ATM per 100,000 adults is associated with fewer number of bank branches per 100,000 adults for G7 countries. On the other hand, the EP and CD coefficient (−0.139), CD and EG coefficient (−0.033), BR and EG coefficient (−0.105), and the BR and EP coefficient (−0.139) are all statistically insignificant. Therefore, no meaningful conclusion can be drawn for this category.

6 Conclusion

The paper analyzed the association between financial inclusion and environmental sustainability. The two financial inclusion indicators used in the study are the number of ATMs per 100,000 adults and the number of bank branches per 100,000 adults, while the two sustainability indicators used in the study are the environmental policy stringency index and the environmentally adjusted multifactor productivity growth. Pearson correlation statistic was used to test the association between financial inclusion and environmental sustainability.

The main findings show that financial inclusion is positively associated with environmental sustainability particularly among non-EU countries. This means that higher environmental policy stringency, which is a sustainability indicator, is associated with higher ATMs per 100,000 adults, which is a financial inclusion indicator. The findings also reveal a significant and negative association between environmental policy stringency and environmentally adjusted multifactor productivity growth particularly among EU member-countries and European countries. Also, there is a significant and positive association between the number of ATMs per 100,000 adults and the number of bank branches per 100,000 adults particularly in EU member-countries and European countries but not for non-EU countries. More so, there is a significant and negative association between the number of ATMs per 100,000 adults and the number of bank branches per 100,000 adults particularly in non-EU and G7 countries.

The implication of the observed positive association between financial inclusion and environmental sustainability is that financial inclusion objectives and environmental sustainability objectives can complement each other in the drive toward achieving the United Nations sustainable development goals (SDGs). The complementary benefits of financial inclusion and environmental sustainability can help to ensure that formal finance is not only accessible to all members of society but also ensure that users of formal finance use formal financial services in ways that promote the continuity of society and the preservation of the environment. Policy makers should find ways to strengthen existing environmental policies to preserve the environment. Financial authorities, on the other hand, should ensure that formal financial services are delivered in ways that promote sustainable growth in society and the environment. Also, there may be a need to make changes in the national financial inclusion strategy to ensure that financial inclusion is aligned with environmental sustainability goals.

The study has some limitations. One, the sample period used in the study is limited only to a few years. Two, the study is restricted only to OECD countries. Three, the study uses a narrow set of financial inclusion and environmental sustainability indictors. This was largely due to data unavailability problems. These limitations offer some fruitful avenue for future research.

Future studies can re-examine the association between financial inclusion and environmental sustainability in other regional contexts such as in the MENA region and the ASEAN region. Secondly, it will be interesting to investigate the cause-and-effect relationship between financial inclusion and environmental sustainability while taking into account any potential endogeneity issues that may arise from such investigation. Finally, future studies can examine the association between financial inclusion and environmental sustainability using more recent data when such data becomes available.

References

Ahmed K, Ahmed S (2018) A predictive analysis of CO_2 emissions, environmental policy stringency, and economic growth in China. Environ Sci Pollut Res 25(16):16091–16100
Albrizio S, Koźluk T, Zipperer V (2014) Empirical evidence on the effects of environmental policy stringency on productivity growth
Arner DW, Buckley RP, Zetzsche DA, Veidt R (2020) Sustainability, FinTech and financial inclusion. European Business Organization Law Review 1–29
Chibba M (2009) Financial inclusion, poverty reduction and the millennium development goals. The European Journal of Development Research 21(2):213–230
Datta SK, Singh K (2019) Variation and determinants of financial inclusion and their association with human development: a cross-country analysis. IIMB Manag Rev 31(4):336–349
De Matteis L (2015) Financial inclusion, policies and instruments for migrants in Europe and Italy. Savings and Development, Special Issue, pp 39–69
Didenko IV, Paucz-Olszewska J, Leonov SL, Ostrowska-Dankiewicz A, Ciekanowski Z (2020) Social safety and behavioral aspects of populations financial inclusion: a multicountry analysis. Journal of International Studies 13(2):347–359
Galeotti M, Salini S, Verdolini E (2020) Measuring environmental policy stringency: approaches, validity, and impact on environmental innovation and energy efficiency. Energy Policy 136:111052
Gu W, Hussain J, Willox M (2019) Environmentally adjusted multifactor productivity growth for the Canadian manufacturing sector. Statistics Canada
Gujarati DN, Porter DC (1999) Essentials of econometrics, vol 2. Irwin/McGraw-Hill, Singapore
Haque F, Ntim CG (2018) Environmental policy, sustainable development, governance mechanisms and environmental performance. Bus Strateg Environ 27(3):415–435
Johnstone N, Haščič I, Poirier J, Hemar M, Michel C (2012) Environmental policy stringency and technological innovation: evidence from survey data and patent counts. Appl Econ 44(17):2157–2170
Kartick D (2012) Financial inclusion-a gateway to sustainable development for the impoverished. J Rural Dev 31(1):115–128
Klapper L, El-Zoghbi M, Hess J (2016) Achieving the sustainable development goals. The role of financial inclusion. Available online: https://www.ccgap.org. Accessed 23(5), 2016
Le Q, Ho H, Mai N (2019a) The impact of financial inclusion on income inequality in transition economies. Management Science Letters 9(5):661–672
Le TH, Chuc AT, Taghizadeh-Hesary F (2019b) Financial inclusion and its impact on financial efficiency and sustainability: empirical evidence from Asia. Borsa Istanbul Rev 19(4):310–322
Lenka SK, Bairwa AK (2016) Does financial inclusion affect monetary policy in SAARC countries? Cogent Economics and Finance 4(1):1127011
Lewis S, Lindley D (2015) Financial inclusion, financial education, and financial regulation in the United Kingdom. October 12. ADBI Working Paper 544
Machdar NM (2020) Financial inclusion, financial stability and sustainability in the banking sector: the case of Indonesia. Int J Econ Bus Adm 8(1):193–202
Martínez-Zarzoso I, Bengochea-Morancho A, Morales-Lage R (2019) Does environmental policy stringency foster innovation and productivity in OECD countries? Energy Policy 134:110982
Morales-Lage R, Bengochea Morancho A, Martínez-Zarzoso I (2016) Does environmental policy stringency foster innovation and productivity in OECD countries? Available at SSRN 2764834
Naumenkova S, Mishchenko S, Dorofeiev D (2019) Digital financial inclusion: evidence from Ukraine. Investment Management and Financial Innovations 16(3):194–205
Neaime S, Gaysset I (2018) Financial inclusion and stability in MENA: evidence from poverty and inequality. Financ Res Lett 24:230–237
Neuberger D (2015) Financial inclusion, regulation, and education in Germany. ADBI Working Paper 530

Ozili PK (2018) Impact of digital finance on financial inclusion and stability. Borsa Istanbul Rev 18(4):329–340
Ozili PK (2020a) Theories of financial inclusion. In: Uncertainty and challenges in contemporary economic behaviour. Emerald Publishing Limited
Ozili PK (2020b) Social inclusion and financial inclusion: international evidence. International Journal of Development Issues 19(2):169–186
Ozili PK (2021a) Has financial inclusion made the financial sector riskier? Journal of Financial Regulation and Compliance 29(3):237–255
Ozili PK (2021b) Financial inclusion research around the world: a review. Forum for Social Economics 50(4):457–479
Park CY, Mercado R Jr (2018) Financial inclusion, poverty, and income inequality. The Singapore Economic Review 63(01):185–206
Piaw CY (2013) Mastering research statistics. Malaysia: McGraw Hill Education, New York, United States
Rahim N, Wan K, Franceschelli M (2009) Financial inclusion amongst new migrants in Northern Ireland: a literature review. Information Centre about Asylum and Refugees, London
Raza MS, Tang J, Rubab S, Wen X (2019) Determining the nexus between financial inclusion and economic development in Pakistan. Journal of Money Laundering Control 22(2):195–209
Rodríguez MC, Haščič I, Souchier M (2018) Environmentally adjusted multifactor productivity: methodology and empirical results for OECD and G20 countries. Ecol Econ 153:147–160
Taddese Bekele D, Abebaw Degu A (2021) The effect of financial sector development on economic growth of selected sub-Saharan Africa countries. International Journal of Finance and Economics
Tolliver C, Fujii H, Keeley AR, Managi S (2021) Green innovation and finance in Asia. Asian Economic Policy Review e12320
Van DTT, Linh NH (2019) The impacts of financial inclusion on economic development: cases in Asian-Pacific countries. Comparative Economic Research 22(1):7–16
Van LTH, Vo AT, Nguyen NT, Vo DH (2021) Financial inclusion and economic growth: an international evidence. Emerg Mark Financ Trade 57(1):239–263
Voica MC (2017) Financial inclusion as a tool for sustainable development
Yin X, Xu X, Chen Q, Peng J (2019) The sustainable development of financial inclusion: how can monetary policy and economic fundamental interact with it effectively? Sustainability 11(9):2524
Yuan B, Zhang Y (2020) Flexible environmental policy, technological innovation and sustainable development of China's industry: the moderating effect of environment regulatory enforcement. J Clean Prod 243:118543

Ozili PK (2018) Impact of digital finance on financial inclusion and stability. Borsa Istanbul Review

eNaira Central Bank Digital Currency (CBDC) for Financial Inclusion in Nigeria

Peterson K. Ozili

1 Introduction

This paper explores the eNaira CBDC and how it can increase financial inclusion in Nigeria. Central bank digital currency is defined as fiat digital money (Jia 2020; Lee et al. 2021; Inozemtsev and Nektov 2022). It is the digital equivalent of physical money (Cunha et al. 2021; Bordo and Levin 2017; Barontini and Holden 2019).

Recently, many central banks have expressed interest in studying about central bank digital currency. Central banks are interested in CBDCs for different reasons. Some central banks plan to issue a CBDC for the purpose of broadening financial inclusion and to improve payment efficiency (Zhang and Huang 2022; Sethaput and Innet 2021; Calle and Eidan 2020). Other central banks plan to issue a CBDC to enhance the conduct of monetary policy, control banking system liquidity, improve payment efficiency, increase government revenue and to provide a legal tender that is relevant to the digital payment ecosystem (Piazzesi and Schneider 2020; Davoodalhosseini et al. 2020; Ozili 2022a). Few central banks have launched a CBDC, notably the Bahamas, China, and Nigeria. Other central banks have launched a CBDC pilot test, while a large majority of central banks in developed countries are still studying about CBDCs to assess whether there is a need for a national CBDC (Ozili 2022a).

There are many benefits of issuing a CBDC. Arguably, the most important benefit of issuing a CBDC in developing countries is to increase financial inclusion since many developing countries have a large unbanked population (Banet and Lebeau 2022; Ozili 2022b). Financial inclusion ensures that all individuals and businesses have access to the basic financial services they need from the formal financial sector (Liu and Walheer 2022; Liu et al. 2022; Ozili 2021). In Nigeria, financial exclusion is an important issue, and the central bank of Nigeria believes that the eNaira CBDC

P. K. Ozili (✉)
Governors Department, Central Bank of Nigeria, Abuja, Nigeria
e-mail: petersonkitakogelu@yahoo.com

M. El Amine Abdelli and M. Shahbaz (eds.), *Digital Economy, Energy and Sustainability*, Green Energy and Technology, https://doi.org/10.1007/978-3-031-22382-2_3

can play a significant role in broadening financial inclusion for the benefit of all Nigerians. In this paper, I show how eNaira can increase financial inclusion.

This study contributes to the CBDC literature that investigate the potential benefits of a central bank digital currency (Baronchelli et al. 2022; Hayashi and Toh 2022; Ozili 2020; Dakila 2022; Maniff 2020; Babin et al. 2022). In the literature, financial inclusion is widely acknowledged to be a major reason for adopting a CBDC in developing countries but it is not a reason for issuing a CBDC in developed economies (Maniff 2020). The discussion in the article also add to the existing literature that investigates the relevance of CBDC in developing countries.

The remainder of the paper is organized in the following way. The literature review is shown in Sect. 2. Analysis of trend data is presented in Sect. 3. An overview of the Nigeria eNaira CBDC is presented in Sect. 4. Section 5 identifies how the eNaira central bank digital currency can increase financial inclusion. Section 6 suggests some important considerations for the future. Section 7 concludes.

2 Literature Review

2.1 Empirical Studies

Ozili (2022d) argued in support of the view that central bank digital currency can increase financial inclusion in Nigeria. The author argued that central bank digital currencies will lead to the effective digitization of value chains in Nigeria, it will improve access to essential digital financial services, it will enlarge the growing digital economy in Nigeria, it will enhance the efficiency of digital payments, and it will offer low transaction costs for users of CBDC in Nigeria.

Maniff (2020) argues that central bank digital currency may experience some difficulty in increasing financial inclusion if the design of the central bank digital currency conflict with the other objectives for issuing a central bank digital currency. The author also argues that a central bank digital currency that is created for financial inclusion purposes would complement cash. The author further points out that central bank digital currency can increase the level of financial inclusion by modernizing payment systems through new, more efficient technologies that provide additional functionality. A central bank digital currency can also increase financial inclusion by offering a viable solution to the frictions in cross-border (international) payments. A central bank digital currency can also provide a public sector solution that delivers cheaper, more efficient cross-border payments. Murakami et al. (2022) argue that central bank digital currencies present some advantages. Some of the advantages include increased financial inclusion of underserved adults and the unbanked population, improving cross-border payments, and facilitating fiscal transfers. However, they emphasize that there are still many unresolved issues in the design of a central bank digital currency. They use a two-agent framework, and show that central bank digital currencies can address financial inclusion of the unbanked if households use CBDC as a savings vehicle to smooth consumption. They also show that retail central bank digital currencies are more useful and beneficial in economies that have low levels of financial inclusion. Engert and Fung (2017) suggest that even though central bank digital currency might improve financial inclusion, financial inclusion is not a major problem in most advanced economies, and for this reason, financial inclusion is not a compelling motivation for issuing a central bank digital currency in most advanced economies including Canada. They emphasize that financial inclusion is most likely an important concern in some emerging and developing economies. Armas et al. (2022) explore the potential role of CBDC implementation in promoting financial inclusion in Peru. They argue that central bank digital currency can increase financial inclusion by allowing the unbanked population to access digital payment instruments. They argue that this can happen by (i) funds transfers where the banking network is not present, (ii) payments on public transportation, (iii) payment of wages in the informal sector and rural areas, (iv) payments to small merchants' suppliers (v) programs to promote social inclusion among the unbanked population (government to person), and (vi) extend accessibility of digital payments to the government. However, they show that the development of the domestic payment system is limited by low financial inclusion. Barr et al. (2020) argue that central bank digital currency can increase financial inclusion by expanding access to financial services using a central bank digital currency. Allen et al. (2022) point out that central bank digital currency will increase financial inclusion by reducing transaction fees and the cost of financial services, thereby improving access to capital. Ozili (2022b) argues that central bank digital currency can increase financial inclusion because consumers do not need to own a bank account to hold CBDC. Ozili (2022a), in a review of literature, shows that the need to increase the level of financial inclusion is a motivation for issuing a central bank digital currency in emerging and developing countries. The author argues that central bank digital currency can increase financial inclusion by giving people access to payments through central bank digital currency. Morales-Resendiz et al. (2021) show that financial inclusion is a major concern in Latin America and the Caribbean (LAC) countries, and the issuance of a central bank digital currency is being considered in these countries to expand access to financial services in Latin America and the Caribbean (LAC) countries. Banet and Lebeau (2022) analyze the impact of introducing a central bank digital currency on financial inclusion. They acknowledge that there has been a growing interest in investigating the potential of a central bank digital currency to support financial inclusion because of the low-cost payments it offers. Gopane (2019) shows that one advantage of central bank digital currency is to broaden financial inclusion and that central bank digital currency is positively correlated with financial inclusion but is negatively affected by digital inequalities among various segments of the population, especially older people. Zuluaga (2021) argues that central banks always cite financial inclusion as a justification for issuing a central bank digital currency because it helps to reduce high fees which make them very appealing to unbanked adults. But the success of CBDC for financial inclusion will depend on whether central banks will rely on private firms to competitively deliver user interfaces such as digital wallets, mobile applications, and customer service. Zuluaga (2021) further argues that even if central bank digital currency becomes attractive to many unbanked households, there will be some unbanked adults whose needs CBDCs cannot satisfy. Zuluaga (2021) emphasizes that even though technology is helping to increase financial inclusion through CBDC, progress in this area is not automatic and it may take many years to achieve. Maryaningsih et al. (2022) show that retail CBDC is more prevalent in countries that have low level of financial inclusion and a large informal economy because it offers easier access to the financial system. They consider financial inclusion to be a determinant of CBDC adoption. Negrea and Scarlat (2022) point out that CBDC issuance can help to increase financial inclusion and create a very competitive market for private-sector payment solutions. In addition, CBDC can create a digital solution that can help to increase financial inclusion for the 1.7 billion people globally who are not banked or do not have access to essential financial services. CBDC can also be used to serve unbanked adults living in areas where traditional banking or financial infrastructure does not exist. Therefore, CBDC can help to achieve a meaning level of financial inclusion in such areas.

2.2 *Theories*

Several theories have emerged that explain the reason for using instruments or tools to achieve high levels of financial inclusion. The most notable theories are the '*theories of financial inclusion*' developed by Ozili (2020). For instance, the public good theory of financial inclusion argues that the state should treat 'financial inclusion' as a public good that should benefit everybody. When this happens, the government will use any relevant tool at its disposal to increase the level of financial inclusion as a public good. The government has a compelling reason to use one of its tools, such as a CBDC, to broaden financial inclusion for the benefit of all citizens so that nobody is left out. The government may also have to subsidize the cost of offering financial services through CBDC payments to encourage more people to own a CBDC account for greater financial inclusion.

Another theory is the dissatisfaction theory of financial inclusion. The theory proposes that financial exclusion occurs when people become dissatisfied with the actions of financial institutions to their customers. The customers become frustrated, and their frustration lead them to exit the financial system, preferring to return to the informal sector because they have lost confidence in financial institutions and can no longer trust financial institutions. The affected unbanked individuals would rather trust a central bank. A central bank can take advantage of people's dissatisfaction with financial institutions by providing them with another alternative way to access the financial system without dealing directly with financial institutions. One such alternative way is through the issuance of a CBDC. Another theory is the public money theory of financial inclusion which emphasizes that public money (e.g. a CBDC) should be used to fund financial inclusion programs.

3 Analyzing Interest Over Time in eNaira and Financial Inclusion in Nigeria

3.1 Local Interest Over Time in eNaira and Financial Inclusion

Interest over time data for Nigeria were collected from Google Trends database from September 25 to October 10 2022. The interest over time data measures the popularity of the 'eNaira' and 'financial inclusion' on the Internet. The numbers represent search interest relative to the highest point on the chart for the given region and time. A value of 100 is the peak popularity for the term. A value of 50 means that the term is half as popular. A score of 0 means there was not enough data for this term. Table 1 shows that the highest interest in eNaira on the Internet was recorded on the day it was launched, while interest in financial inclusion reached its highest point in May 2022. Table 2 shows that northern states in Nigeria had greater interest in eNaira than southern states particularly Gombe State, Adamawa State, Jigawa State, and Yobe State (Figs. 1 and 2).

4 The eNaira

Nigeria launched its own fiat CBDC commonly known as the 'eNaira' on the September 25, 2021. Nigeria launched the eNaira at a time when there was heated global debate about the risks of private digital currencies and the need for a government-controlled digital currency (Brunnermeier and Niepelt 2019; Ozili 2022c). Prior to launching the eNaira CBDC, the central bank of Nigeria, in February 2021, barred regulated deposit money banks from aiding or facilitating cryptocurrency transactions because fraudsters used cryptocurrencies to hide illicit financial dealings (Ozili 2022d). Such practices undermined the integrity of the Nigerian banking system. The implication of the restriction is that cryptocurrency transactions are no longer permitted in the Nigerian banking system. Soon after the restriction was imposed, the central bank launched the eNaira central bank digital currency. The eNaira CBDC has the same features of the paper Naira, and it offers more payment possibilities for Nigerians.[1] The eNaira CBDC is designed to act as an efficient payment tool. It will compete with existing payment channels. The eNaira also offers many benefits. The eNaira helps to (i) increase financial inclusion, (ii) enhance the efficiency of payments, (iii) encourage broad digitization in the Nigerian society, (iv) increase cross-border trade, (v) facilitate cheaper and faster remittance inflows, (vi) implement targeted social interventions, (vii) improve monetary policy effectiveness, and (viii) improve tax collection.

---

[1] Extensive details about the eNaira CBDC design features and structure can be found in the article: Ozili (2022d).

Table 1 Interest over time data

Week	eNaira: (Nigeria)	Financial inclusion: (Nigeria)
31/10/2021	100	0
07/11/2021	76	79
14/11/2021	48	38
21/11/2021	31	0
28/11/2021	24	39
05/12/2021	43	56
12/12/2021	22	0
19/12/2021	33	0
26/12/2021	30	0
02/01/2022	20	65
09/01/2022	10	80
16/01/2022	8	74
23/01/2022	23	68
30/01/2022	18	60
06/02/2022	15	54
13/02/2022	21	22
20/02/2022	11	27
27/02/2022	9	19
06/03/2022	12	51
13/03/2022	13	46
20/03/2022	12	26
27/03/2022	3	48
03/04/2022	9	0
10/04/2022	5	0
17/04/2022	6	53
24/04/2022	0	57
01/05/2022	5	20
08/05/2022	5	100
15/05/2022	14	0
22/05/2022	15	60
29/05/2022	8	42
05/06/2022	6	32
12/06/2022	14	44
19/06/2022	0	75
26/06/2022	5	0
03/07/2022	5	74
10/07/2022	13	61

(continued)

Table 1 (continued)

Week	eNaira: (Nigeria)	Financial inclusion: (Nigeria)
17/07/2022	9	47
24/07/2022	7	31
31/07/2022	8	22
07/08/2022	16	0
14/08/2022	16	0
21/08/2022	20	64
28/08/2022	14	62
04/09/2022	36	0
11/09/2022	16	20
18/09/2022	22	0
25/09/2022	17	0
02/10/2022	10	24

5 eNaira for Financial Inclusion

One important benefit of the eNaira CBDC is that it can increase financial inclusion in Nigeria in several ways. Below are examples of how the eNaira can increase financial inclusion in Nigeria.

- **eNaira offers an easy account opening process for greater financial inclusion**
  The eNaira central bank digital currency has a simplified account opening process. Individuals do not need to produce any burdensome documentation to open an eNaira account. The easy account opening process will motivate more citizens to open an eNaira account to avoid the burdensome account opening documentation required by the deposit money banks.
- **eNaira enables digital access to diverse financial services in the financial system**
  The eNaira central bank digital currency will be accessible to Nigerians through the eNaira speed wallet and a USSD code so that they can access the eNaira CBDC both online or offline. With the eNaira, Nigerians will be able to access available digital financial services at the comfort of their homes without the hassle of visiting a bank branch to make or receive payments. They will also be able to make eNaira payments at check-out when making purchases.
- **eNaira offers lower cost of financial products and services**
  The eNaira offers very low transaction cost when making payments. This will make the eNaira a cheaper payment alternative compared to existing payment channels. The eNaira will also help to reduce the cost of remittance and make it cheaper for Nigeria diaspora to send money back home by obtaining eNaira from international money transfer operators (IMTOs) and transferring the remitted funds to the recipients or beneficiaries living in Nigeria by wallet-to-wallet transfers free of charge.

Table 2 Interest over time according to region

Region	eNaira: (10/25/21–10/10/22)
Gombe	100
Adamawa	99
Jigawa	95
Yobe	91
Zamfara	85
Katsina	75
Borno	74
Kano	74
Kebbi	73
Niger	73
Kaduna	50
Federal Capital Territory	42
Plateau	39
Bauchi	38
Rivers	37
Oyo	35
Abia	34
Kwara	31
Ogun State	31
Ebonyi	29
Osun	27
Benue	26
Sokoto	25
Nasarawa	25
Ondo	24
Lagos	23
Anambra	22
Akwa Ibom	20
Edo	19
Delta	14
Ekiti	13
Cross River	13
Imo	10
Kogi	10
Enugu	5

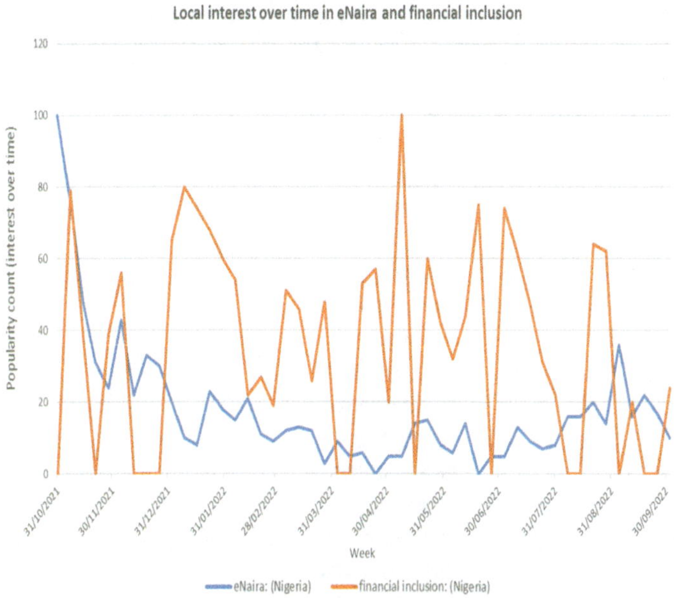

Fig. 1 Local interest over time in eNaira and financial inclusion

- **No more unexplained bank charges that causes financial exclusion**
  One reason why people voluntarily exit the formal financial system is because of unexplained bank charges. The charges affect the poor severely, and they react by exiting the formal financial system. The introduction of the eNaira will bring an end to unexplained bank charges. Users of eNaira will pay very low transaction cost, and they will understand the purpose of the transaction cost. They will be clearly informed and notified about a change in transaction cost or the introduction of new charges. This will increase central bank transparency, increase trust in the government, and make more unbanked adults willing to join to formal financial system by opening an eNaira account.
- **eNaira will be used to attract people who have lost confidence in banks**
  People may lose confidence in commercial banks when they become victims of fraud or victims of unauthorized access to bank account that lead to loss of their savings. Such experience can make people exit the financial system due to dissatisfaction (Ozili 2020).[2] The eNaira can accommodate the needs of this category of people. People who have lost confidence in commercial banks will trust the eNaira since the eNaira is a direct liability of the central bank not commercial banks. They can be confident that if they keep their money with the central bank through the eNaira, their savings will be protected and secured, and the central

[2] This is consistent with the propositions of the 'dissatisfaction theory of financial inclusion' A proposed by Peterson Ozili in his work 'Theories of financial inclusion'.

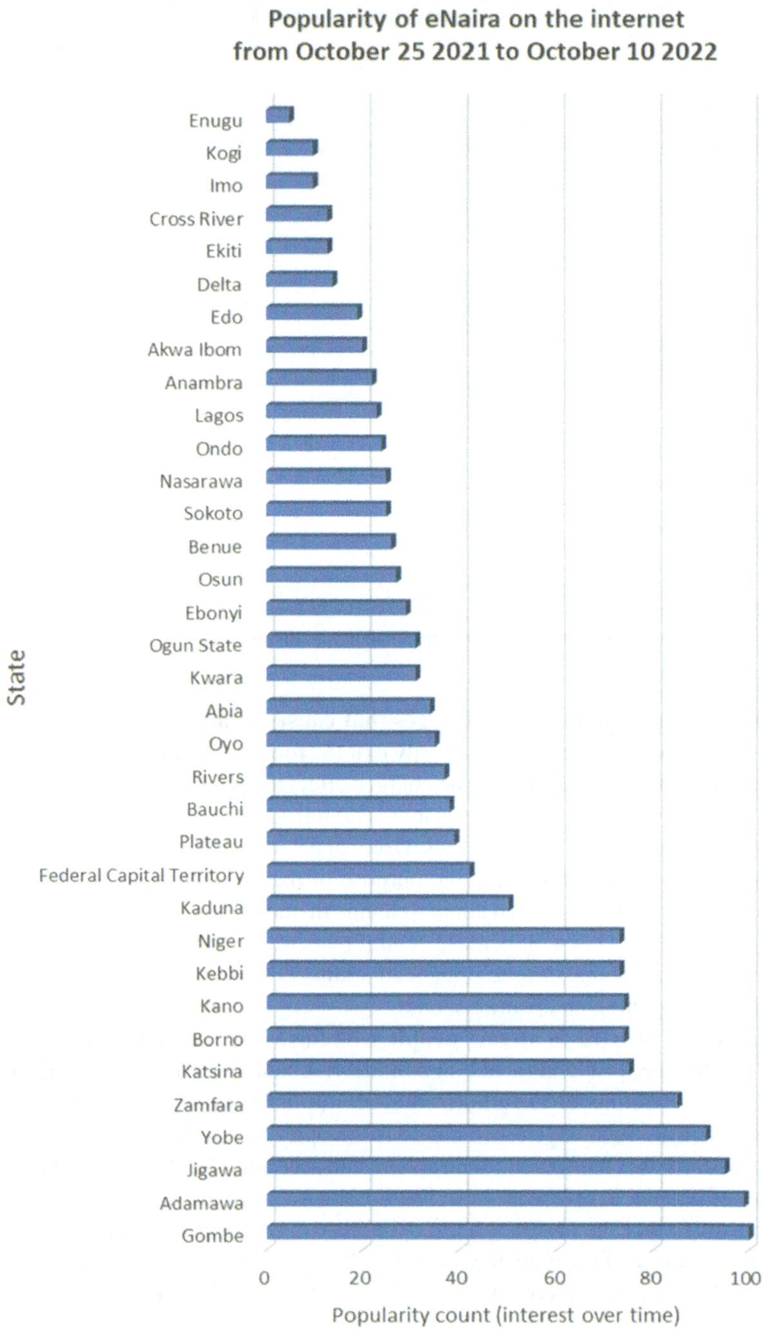

Fig. 2 Popularity of eNaira on the internet

bank will bear liability for any loss of customers' money that is not the customer's fault.

- **Interest-bearing eNaira CBDC can increase financial inclusion**
  An interest-bearing eNaira CBDC will increase financial inclusion. The interest earned on eNaira deposits will be attractive to unbanked adults and will give unbanked adults a motivation to join the financial system through the eNaira. They will bring their savings to the formal financial system through the eNaira. They will also have access to wide range of financial services that can be accessed through the eNaira.
- **eNaira can increase financial inclusion using offline channels**
  The eNaira can increase financial inclusion by using a USSD code to onboard the unbanked and underserved users that do not have access to Internet connectivity or a smartphone.

6 Important Considerations for the Future

Some important considerations to increase the potential of the eNaira to increase financial inclusion include the following.

- **eNaira adoption should be gradual.** Enough time should be given to citizens and businesses to adopt the eNaira CBDC. This is important because it will ensure that citizens and businesses use the eNaira because there is a clear benefit of using the eNaira and not because they are compelled to adopt it. People should be allowed to compare the benefits of the eNaira with the benefits of other payment alternative. This will help them to choose the eNaira if it offers superior benefits.
- **Widespread digital literacy among the rural population is essential for greater eNaira adoption and usage.** There is significant digital illiteracy in some communities in Nigeria. Digital illiteracy is highly undesirable in today's society. It is a cause of digital exclusion and social exclusion. Digital illiteracy affects vulnerable people, especially those living in rural and remote areas. When people are unable to use digital tools on a daily basis, it prevents them from accessing basic resources that can only be accessed digitally. They won't be able to access financial services remotely because they don't know how to, and they won't be able to teach their family and peers how to use digital tools. Some causes of digital illiteracy in Nigeria are lack of education, improper education, lack of access to Internet connectivity, and lack of access to mobile phones (Ekoh et al. 2021; Ani et al. 2007). The rural population needs basic digital literacy to enable them to fully integrate in the digital society (Tayo et al. 2016). They should learn how to download an app, update an app, login into an app, transfer money on the app, and check transaction history on the app. They should also learn how to use a USSD code to perform financial transactions.
- **The eNaira should not replace cash completely**. The eNaira CBDC should co-exist with cash because cash remains the dominant form of transaction in Nigeria,

especially among the financially excluded groups. People need time to adjust to using the eNaira CBDC.

- **eNaira deposits should be interest-bearing to attract unbanked savers**. Unbanked savers will be motivated to bring their savings into the formal financial system through the eNaira because they trust the central bank. The interest earned on eNaira deposits should be lower than the interest on bank deposits to avoid disorderly bank disintermediation.
- **The eNaira CBDC should be non-disruptive**. The phased introduction of eNaira CBDC should not significantly disrupt the financial and banking system.
- **The eNaira may not be well-perceived or welcomed in the informal economy**. The central bank needs to find a way to make the eNaira accepted in the informal economy in Nigeria.

7 Conclusion

This paper explored how eNaira CBDC can achieve financial inclusion in Nigeria. The eNaira is being used to broaden financial inclusion as part of the National Financial Inclusion Strategy. It was argued that the eNaira can increase financial inclusion by (i) offering an easy account opening process for greater financial inclusion, (ii) enabling digital access to diverse financial services in the financial system, (iii) offering lower cost of financial products and services, (iv) avoiding unexplained bank charges that cause financial exclusion, (v) attracting people who have lost confidence in banks, (vi) introducing interest-bearing eNaira, and (vii) using offline channels to access the eNaira.

The implication of the findings is that there is a clear channel through which eNaira can increase financial inclusion. Policymakers should explore other innovative and non-disruptive ways through which the eNaira CBDC can increase financial inclusion for the good of Nigerians.

Although there is a clear pass-through between eNaira usage and financial inclusion, a number of considerations need to be taken into account. Some of which include embarking on a national digital literacy program to increase awareness of the eNaira. There is also a need to watch out for risks related to cyber-attacks and user protection.

Future studies should examine how the eNaira affects economic growth. Such studies should focus on how the eNaira affects business activities in specific sectors of the Nigerian economy. Future studies can also compare the Nigeria eNaira CBDC with other CBDC in Africa such as the Ghana eCedi. Such comparison will provide insight in understanding the local modifications in CBDC design.

References

Allen F, Gu X, Jagtiani J (2022) Fintech, cryptocurrencies, and CBDC: financial structural transformation in China. Journal of International Money and Finance 124:102625

Ani OE, Uchendu C, Atseye EU (2007) Bridging the digital divide in Nigeria: a study of internet use in Calabar Metropolis, Nigeria. Library Management.

Armas A, Ruiz L, Vásquez JL (2022) Assessing CBDC potential for developing payment systems and promoting financial inclusion in Peru. BIS Papers Chapters 123:131–151

Babin R, Smith D, Shah H (2022) Central bank digital currency: advising the financial services industry. Journal of Information Technology Teaching Cases 20438869221116901

Banet J, Lebeau L (2022) Central bank digital currency: financial inclusion vs. disintermediation. Federal Reserve Bank of Dallas, Working Paper 2218

Baronchelli A, Halaburda H, Teytelboym A (2022) Central bank digital currencies risk becoming a digital Leviathan. Nat Hum Behav 6(7):907–909

Barontini C, Holden H (2019) Proceeding with caution-a survey on central bank digital currency. In: Proceeding with caution-a survey on central bank digital currency. BIS Paper, No. 101, 8 Jan 2019

Barr MS, Harris A, Menand L, Xu WM (2020) Building the payment system of the future: how central banks can improve payments to enhance financial inclusion. U of Michigan Law & Econ Research Paper (20-038)

Bordo MD, Levin AT (2017) Central bank digital currency and the future of monetary policy. National Bureau of Economic Research, No. 23711

Brunnermeier MK, Niepelt D (2019) On the equivalence of private and public money. J Monet Econ 106:27–41

Calle G, Eidan D (2020) Central bank digital currency: an innovation in payments. R3 White Paper, April

Cunha PR, Melo P, Sebastião H (2021) From bitcoin to central bank digital currencies: making sense of the digital money revolution. Future Internet 13(7):165

Dakila FG Jr (2022) Deliberations of an emerging market economy central bank on central bank digital currencies. BIS Papers Chapters 123:153–160

Davoodalhosseini M, Rivadeneyra F, Zhu Y (2020) CBDC and monetary policy (No. 2020-4). Bank of Canada

Ekoh PC, George EO, Ezulike CD (2021) Digital and physical social exclusion of older people in rural Nigeria in the time of COVID-19. J Gerontol Soc Work 64(6):629–642

Engert W, Fung BSC (2017) Central bank digital currency: motivations and implications (No. 2017-16). Bank of Canada Staff Discussion Paper

Gopane TJ (2019) An enquiry into digital inequality implications for central bank digital currency. In: 2019 IST-Africa week conference (IST-Africa). IEEE, pp 1–9

Hayashi F, Toh YL (2022) Assessing the case for retail CBDCs: central banks' considerations. Payments System Research Briefing, 1–7, 26 May 26, 2022

Inozemtsev MI, Nektov AV (2022) Digital platforms for cross-border settlement of CBDC. In: The platform economy. Palgrave Macmillan, Singapore, pp 131–145

Jia P (2020) Negative interest rates on central bank digital currency

Lee DKC, Yan L, Wang Y (2021) A global perspective on central bank digital currency. China Economic Journal 14(1):52–66

Liu F, Walheer B (2022) Financial inclusion, financial technology, and economic development: a composite index approach. Empirical Economics, 1–31

Liu D, Xie Y, Hafeez M, Usman A (2022) The trade-off between economic performance and environmental quality: does financial inclusion matter for emerging Asian economies? Environ Sci Pollut Res 29(20):29746–29755

Maniff JL (2020) Motives matter: examining potential tension in central bank digital currency designs. Federal Reserve Bank of Kansas City: Omaha, NE, USA

Maryaningsih N, Nazara S, Kacaribu FN, Juhro SM (2022) Central bank digital currency: what factors determine its adoption? Buletin Ekonomi Moneter Dan Perbankan 25(1):1–24
Morales-Resendiz R, Ponce J, Picardo P, Velasco A, Chen B, Sanz L, Guiborg G, Segendorff B, Vasquez JL, Arroyo J, Aguirre I (2021) Implementing a retail CBDC: lessons learned and key insights. Latin American Journal of Central Banking 2(1):100022
Murakami D, Shchapov I, Viswanath-Natraj G (2022) CBDCs, financial inclusion, and monetary policy. Available at SSRN 4102397
Negrea CI, Scarlat EM (2022) The digital Leu challenges and possible areas of implementation. "Ovidius" University Annals, Economic Sciences Series, vol XXII, Issue 1
Ozili PK (2020) Theories of financial inclusion. In: Uncertainty and challenges in contemporary economic behaviour. Emerald Publishing Limited
Ozili PK (2021) Financial inclusion research around the world: a review. Forum for Social Economics 50(4):457–479
Ozili PK (2022a) Central bank digital currency research around the World: a review of literature. Journal of Money Laundering Control
Ozili PK (2022b) Can central bank digital currency increase financial inclusion? Arguments for and against. In: Big data analytics in the insurance market. Emerald Publishing Limited, pp 241–249
Ozili PK (2022c) Global central bank digital currency research and developments: implication for cryptocurrency. Cryptocurrency: Concepts, Technology, and Issues, Forthcoming
Ozili PK (2022d) Central bank digital currency in Nigeria: opportunities and risks. In: The new digital era: digitalisation, emerging risks and opportunities, vol 109. Emerald Publishing Limited, pp 125–133
Piazzesi M, Schneider M (2020) Credit lines, bank deposits or CBDC? Competition and efficiency in modern payment systems. Stanford University (unpublished)
Sethaput V, Innet S (2021) Blockchain application for central bank digital currencies (CBDC). In: 2021 third international conference on blockchain computing and applications (BCCA). IEEE, pp 3–10
Tayo O, Thompson R, Thompson E (2016) Impact of the digital divide on computer use and internet access on the poor in Nigeria. Journal of Education and Learning 5(1):1–6
Zhang T, Huang Z (2022) Blockchain and central bank digital currency. ICT Express 8(2):264–270
Zuluaga D (2021) Which type of digital currency for financial inclusion? Cato J 41:413

Catalyzing Climate Finance for Climate Actions in MENA Countries: A Holistic View of Egypt, Morocco, and Tunisia

Sana Essaber, Abdelaziz Essayem, and Imen Baccouche

1 Introduction

The Paris Agreement clearly states that all funding sources must achieve the adaptation and mitigation goals set out in the agreement. Achieving these goals depends on developing countries committing to emission reduction programs to drive low-carbon development and adaptation programs for more climate-resilient countries. The developing countries indicate their Defined National Contributions (NDCs) actions with local resources and the potential actions to be taken if they receive international support. They commit to reducing their greenhouse gas (GHG) emissions voluntarily. In return, developed countries commit to the principle of common but differentiated responsibility. This requires massive investments in new infrastructure and bringing existing ones into line with the environment.

Moreover, current financing and investment patterns are considered unable to deliver sustainable development. In particular, expected returns on investments associated with sustainable development are often not as attractive as other opportunities, especially in the near term. At the same time, there are many competing demands on public resources, and governments have not been able to mobilize adequate public financing to undertake necessary investments that profit-seeking investors avoid. Through their survey of green finance literature, Gilchrist and Zhong (2021) argue that developed and developing countries have distinct agendas when establishing

S. Essaber (✉) · A. Essayem
Higher Institute of Accounting and Business Administration (ISCAE), University of Manouba, Manouba, Tunisia
e-mail: sana.essaber@iscae.uma.tn

A. Essayem
MOCFINE, ISCAE, Manouba, Tunisia

S. Essaber · I. Baccouche
LEGI—Tunisia Polytechnic School, University of Carthage, Carthage, Tunisia

M. El Amine Abdelli and M. Shahbaz (eds.), *Digital Economy, Energy and Sustainability*, Green Energy and Technology, https://doi.org/10.1007/978-3-031-22382-2_4

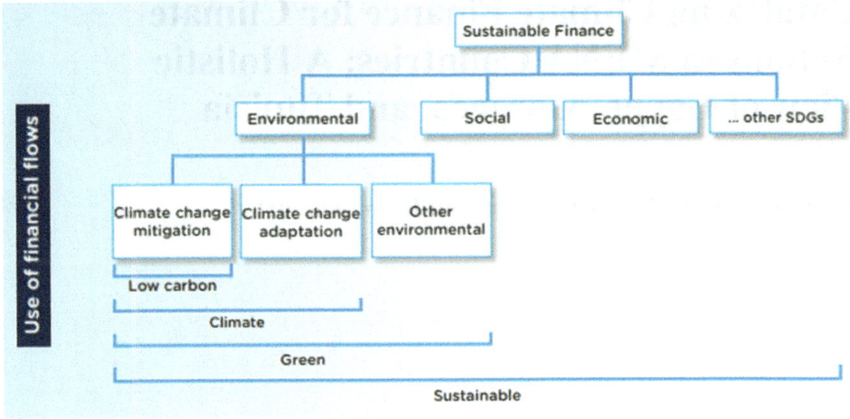

Fig. 1 Elements of sustainable finance

green finance criteria, and presumably, more pressing financial concerns dictate their short-term policy.

Given the importance of the financial resources to be mobilized and the scarce public funding and investments, many public and private initiatives have been undertaken to reduce the climate change impact, so the financial sector can contribute as a conducive vehicle for financing the transition to a climate-friendly economy (Pauw and Pegels 2013; Climate Action Network 2013; OECD 2018). Consequently, climate finance, namely for mitigation and adaptation actions, is one of the green finance dimensions, which in turn are a subset of sustainable development finance (Fig. 1).

The international community and private sector engagement will be very important (Climate Analytics 2019; Climate Policy Initiative 2017). Moreover, aware of the systemic impact of climate risk, the regulator's role has become to support incentives for the financing of the energy transition and put in place an appropriate framework for managing collective risks and financial institutions (Campiglio 2016). The financial sector and all its components (institutional investors, banks, insurance companies, etc.) are particularly exposed to the risks related to climate change (UN Environment 2017). Thus, access to climate finance requires countries to have a clear and concise strategy for planning and implementing climate projects (Buchner et al. 2017).

This research explores the potential sources and new patterns of climate action funds in developing countries. To this end, a detailed and extensive literature survey was performed to explore the barriers and challenges to climate finance in developing countries and better comprehend the core players involved in this area. For illustration, some MENA countries are used as case studies.

The paper is organized as follows: In Sect. 2, we present the principal current climate funding in Egypt, Morocco, and Tunisia. Section 3 treats the main constraints related to climate investment in these MENA countries. We conclude in Sect. 5.

Fig. 2 Sample countries to be studied

2 Current Climate Finance in Egypt, Morocco, and Tunisia

To meet climate change commitments in the future, significant new funds are needed to be provided by various organizations and banks to support MENA governments in promoting innovative strategies, technologies, and financing options to manage the threats related to the climate change impacts and its impact on populations and countries main socio-economic indicators.

Based on the total funds identified in MENA countries, only, a few amounts come from climate funds (GCF,[1] GEF,[2] AF,[3] CTF,[4] FIP[5]). The largest share comes from dedicated climate finance provided by MDBs, in particular, the European Bank for Reconstruction and Development (EBRD), the European Investment Bank (EIB), and the International Bank for Reconstruction and Development (IBRD), and most funding was country-specific, regional projects (Midgley et al. 2018). However, the bulk of the regional funding went to the GCF.

In this work, Egypt, Morocco, and Tunisia (Fig. 2) are retained as study cases to highlight climate funds' financial resources to support climate projects. Climate impact scenarios are generally more serious in MENA than in other regions (Schaar 2019; Cooper 2020).

There are multiple climate funding providers in Egypt, Morocco, and Tunisia through bilateral channels, multilateral channels, or specific climate funds such as the Climate Investment Fund (CIF) and the Green Climate Fund (GCF). The World Bank's Climate Investment (CIF), the Agence Française de Développement (AFD), and the European Bank for Reconstruction and Development (EBRD) all finance

[1] GCF: Green Climate Fund https://www.greenclimate.fund/.

[2] GEF: Global Environment Facility https://www.thegef.org/.

[3] AF: Adaptation Fund https://www.adaptation-fund.org/.

[4] CTF: Clean Technology Fund: https://climatefundsupdate.org/the-funds/clean-technology-fund/.

[5] FIP: Forest Investment Partnership https://www.climateinvestmentfunds.org/sites/cif_enc/files/results-2015/fip/.

Table 1 Climate finance received by the economy from multilateral development banks for 2015, 2016, and 2017 (in US$ million)

Economy	2015	2016	2017	Total
Tunisia	19	96	387	502
Morocco	914	729	668	2310
Egypt	511	693	1585	2789

Source Multilateral Development Banks (2020)

major projects. Tunisia received more from Multilateral Development Banks, 387 million US$ in 2017, less than Morocco and Egypt (Table 1).

Access to Green Climate Fund resources is through accredited national, regional, and international entities to conduct climate change programs and projects. Consistent with the $100 billion pledge goals and the Paris Agreement, mitigation and adaptation goals are prioritized.

However, large investments are much more likely to be directed toward mitigation than adaptation (Pauw 2015) (Table 2). This is due to more established measurement and reporting of mitigation benefits than for adaptation. Renewable energy and transportation projects are also more broadly agreed upon as having mitigation benefits/co-benefits, and these projects often yield a better return on investment.

The CIF's goal is the "*transformational change towards climate-resilient, low-carbon development in developing countries through scaled-up financing.*" CIF's scale and flexible approach help countries mitigate and adapt to climate change in an integrated, programmatic fashion. It was established in 2008 to deploy climate finance at scale to advance clean technology (CTF) and renewable energy (SREP), sustainable management of forests (FIP), and climate-resilient development (PPCR).

The CIF is implemented through five MDBs (i.e., AfDB; ADB; EBRD; IDB; and WBG, including IFC) that strengthen planning and action for climate-friendly development in 72 developing and middle-income countries worldwide.

The Clean Technology Fund (CTF) provides financing for investments in innovative technologies and clean energy projects. The fund provides loans, grants, guarantees, equity, and local currency financing at reduced market rates on a scale that is

Table 2 Climate finance received by the economy from green climate fund (GCF)

	Egypt	Morocco	Tunisia
Number of projects	4	9	6
Total GCF financing (US$ million)	300.7	228.9	120.5
Project nature:			
• Mitigation	2	4	4
• Adaptation	1	2	-
• Cross-cutting	2	3	2

Source Green Climate Fund

Table 3 Climate finance received by the economy from climate investment fund (CIF)

Projects in Egypt				
Name	Fund	Funding (US$ million)	Cofinancing (US$ million)	MDB
DPSP III: Sustainable Urban Infrastructure: Expansion	Clean Technology Fund	30.00	145.00	IFC
Wind Power Development Project	Clean Technology Fund	150.00	653.50	IBRD
Projects in Morocco				
Name	Fund	Funding (US$ million)	Cofinancing (US$ million)	MDB
Clean and Efficient Energy Project	Clean Technology Fund	23.95	134.36	IBRD
One Wind Energy Plan	Clean Technology Fund	125.00	2263.74	AFDB
Projects in Tunisia				
Name	Fund	Project cost (US$ million)	Cofinancing (US$ million)	MDB
Integrated landscape management in Tunisia's lagging regions	FIP	96	–	–
Integration of the tree in degraded private farmland	FIP	34	–	–
Sustainable management of Tunisian rangelands	FIP	28	–	–

Source https://www.climateinvestmentfunds.org/sites/default/files/meeting-documents/fip_investment_plan_for_tunisia_0.pdf

big enough to matter. Egypt and Morocco benefit from the Clean Technology Fund (CTF), and Tunisia benefits from CTF through a regional project but mainly from the forest investment program (FIP). More details are given in Table 3.

3 Constraints Related to Climate Investment in Egypt, Morocco, and Tunisia

The main constraints to achieving climate change mitigation and adaptation projects are institutional, technical, and financial. In addition, climate change-related investments are significantly lower than total foreign direct investment (FDI). This can be justified by the risks of this type of investment, such as exchange rate risk, project liquidity, credit risk, and project operational risks (implementation, the technology used, etc.). The main obstacles to climate investment in MENA countries are recurrent in the rest of the developing countries. We summarize the main constraints for three MENA countries, namely Egypt, Morocco, and Tunisia, in Table 4.

4 Opportunities Related to Climate Investment in Egypt, Morocco, and Tunisia

Various mechanisms to foster a more attractive investment climate have been created at the global, regional, national, and sub-national scales. Developing a national strategy to develop RE in the local market is insufficient. Its implementation requires a favorable regulatory, technical, and economic framework, particularly regarding key issues such as administrative procedures, grid connection permits, or financial incentives (tax rebates, preferential lending rates, etc.). Several factors, such as weak governance, inadequate institutions, imperfect macroeconomic policy and infrastructure, weak legal systems, and lack of public service transparency, have created an adverse investment environment. Addressing these numerous issues may involve new skills, such as financial and risk management expertise, technical knowledge of sustainable development, and skills for evaluating climate projects' financial benefits and sustainability impacts.

Besides, efficient public climate finance could reduce the risks related to climate investment and encourage private funding (Westphal et al. 2015). The central bank's role is essential in enhancing the value of greenhouse gas emission reduction activities and transforming climate finance. Indeed, instituting a monetary value for GHG emission reduction activities could engage more investors. Entrepreneurs who wish to engage in climate investment need some guarantee that entrepreneurs' private gain can also be equivalent to the collective gain for society. However, the market cannot do it by itself. Instead, the Central Bank can introduce this social and monetary value and institute it, which provides a guarantee and a safeguard for these costly investments and long-term returns on investment (Campiglio 2016). The central bank should regulate a deep financial transformation process.

On the one hand, this process requires the certification of the volume of GHGs avoided by the investments made and, on the other hand, the issuance of green bonds. This is the path that may lead to the goal of carbon neutrality (Campiglio 2016). We summarize the key challenges and principal bank activities in Table 5.

Table 4 Main climate projects barriers in Egypt, Morocco, and Tunisia

	Political-regulatory field	Economical-financial field	Technological or other fields
Egypt*	• High risk and uncertainties because of the unstable political situation • The national electricity monopoly • Administrative delaying of bureaucracy • No incentives from the state	• High costs of capital and transactions • High level of interest • Non-involvement of national banks • Opposition to hard currency • Small-size initiatives	• Lack of management capacity • Insufficient local knowledge and expertise • Heavy reliance on fossil fuels (oil and natural gas)
Morocco*	• Initial low level of thinking • No allowances for independent electricity supply • No commitment to transportation emissions • Large subsidies on fuel	• High CDM project fees and costs • The long-time cycle of CDM projects • Debt is not an option for small and medium-sized enterprises (SMEs) • No engagement of domestic banks in the fund-raising decision process • No long-term financing in hard currency • No tax or other financial benefits	• Lack of local expertise in technical matters • Limited consciousness of the issues and mechanisms of the environment • Limited national natural gas reserves
Tunisia**	• The monopoly of STEG is not perceived as positive and may affect public policy • Lack of market visibility • Unclear national vision for the development of a domestic renewable energy market • A confusing set of competition-distorting policies	• The Tunisian currency's inflation and the fact that most investments are made in foreign currency • Absence of a clear and steady implementation framework for renewable energies • Absence of a transparent financing scheme, namely for small and medium-sized enterprises	• Lack of technical specifications for renewable energy developers to be plugged into the electricity grid • Limited expertise in high-value components (photovoltaic cells, float glass, receivers, etc.) • Lack of public as well as private investment in R&D • Lack of methodological tools for climate indicators and project implementation

Sources * Karakosta and Psarras (2013), **Morsy et al. (2018)

Table 5 Main objectives and tasks of the bank

Egypt	• Supporting the competitiveness of the Egyptian private sector by building deeper value networks and chains • Better access to finance for small and medium-sized enterprises (SMEs) • Increasing the number of economic opportunities for both women and youth
Morocco	• Improve access to finance for SMEs • The rural–urban gap results in significant deficits in access to services and finance and significant imbalances in the subjacent business environment • Support sustainability and marketing of public services and infrastructure
Tunisia	• Strengthening financial and economic inclusion of youth and underperforming regions, including support for SMEs, microfinance, and advanced technologies • Funding the growth of green and sustainable transport services (e.g., railroads, ports, logistics zones, intermodal terminals) and energy storage systems • Improve the governance of the financial sector

5 Conclusion

The country's capacity to efficiently manage the climate change risks relies on many elements, including the regulatory and institutional capacity to build and deploy adaptation strategies, the ability to monitor and gather information to respond to the impacts of climate change, the reinforcement of capacities in using technologies dedicated to climate action, as well as access to finance, among others.

Climate change financing sources include all national, regional, or international financial support for activities or projects addressing an aspect of climate change. The study of the Egyptian, Moroccan, and Tunisian cases provides interesting examples of the need for an integrated mitigation and adaptation strategy to access financial resources that are still progressing slowly despite the efforts.

There is not usually sufficient detail on the funding level required to sustain adaptation actions or the potential future benefits/savings from avoided losses that can be realized through resilience and adaptation actions. This last point is related to the fact that, unlike mitigation, there is no single measure to quantify the benefits of adaptation actions. Hence, a need for adaptation finance and concessional finance in the MENA region is identified. Over 80% of climate finance from international public finance flows supports mitigation, largely in renewable energy funding (Midgley et al. 2018; ECSWA 2019); however, adaptation is a key priority among MENA countries (ESCWA 2019).

Despite the availability of financial instruments for climate investments from public and private sources, they are not often accessible due to restrictions on possible beneficiaries, strict eligibility criteria, or lengthy application procedures.

References

Buchner et al (2017) Global landscape of climate finance. Climate Policy Initiative, October

Campiglio E (2016) Beyond carbon pricing: the role of banking and monetary policy in financing the transition to a low-carbon economy. Ecol Econ 121:220–230

Climate Action Network (2013) Climate change adaptation and the role of the private sector: Creating effective tools for private sector engagement, CAN Europe

Climate Analytics (2019) Annual report 2018. Available from: https://climateanalytics.org/publications/2019/climate-analytics-annual-report-2018/

Climate Policy Initiative (2017). Global landscape of climate finance architecture, October. Available from: https://www.climatepolicyinitiative.org/publication/global-landscape-of-climate-finance-2017/

Cooper, R (2020) Climate change risks and opportunities in the Middle East and North Africa, K4D Helpdesk Report, Brighton, UK: Institute of Development Studies

Della et al. (2011), The role of pension funds in financing green growth initiatives, OECD Publishing, Paris, 10

ESCWA—Economic and Social Commission for Western Asia (2019) Technology for development. Annual Report 2018

Gilchrist and Zhong (2021) The limits of green finance: a survey of literature in the context of green bonds and green loans. Sustainability 13:478

Karakosta C, Psarras J (2013) Understanding CDM potential in the Mediterranean basin: A country assessment of Egypt and Morocco, *Energy policy*, 60.

Midgley A et al (2018) International public climate finance in the Mediterranean—Updated results for 2016—Final report. Union for the Mediterranean. Climatekos

Morsy H et al (2018) Tunisia diagnostic paper: assessing progress and challenges in unlocking the private sector's potential and developing a sustainable market economy. European bank for reconstruction and development, November

Multilateral Development Banks (2020) Joint report on multilateral development banks climate finance, August. Available from: https://www.eib.org/attachments/press/1257-joint-report-on-mdbs-climate-finance-2019.pdf

OECD (2018) Climate finance from developed to developing countries: Public flows from 2013–2017. OECD Publications

Pauw P, Pegels A (2013) Private sector engagement in climate change adaptation in least developed countries: an exploration. Climate and Development 5(4), August

Pauw WP (2015) Not a panacea: private-sector engagement in adaptation and adaptation finance in developing countries. Climate Policy 15(5):583–603

Schaar J (2019) A confluence of crises: on water, climate and security in the Middle East and North Africa. SIPRI Insights on Peace and Security

UN Environment (2017) Roadmap for a sustainable financial system, inquiry: design of a sustainable financial system, November

Westphal MI et al (2015) Getting to $100 billion: climate finance scenarios and projections to 2020. World Resources Institute, Working paper, May

Analysis of the Indicators of Environmental Performance in Algeria

Soudani Ahlem and Arabeche Zina

1 Introduction

Algeria is one of the biggest energy-producing and exporting countries in the world. This has been done through the implementation, in an economic, social, and environmental sense, of policies and programs to promote the use of energy resources. Although the policies and strategies implemented have not yet placed the country in its proper position concerning the separation of energy sources. These efforts have not yet made full use of its national renewable and non-renewable energy capacities and reserves to diversify the economy, reduce its dependency on the hydrocarbon industry, and preserve its economic security. Over the last decade, the world price of crude oil has fallen dramatically, and its price has varied to this day, which has seriously affected the country's economy. In addition, the shift to the exploitation of renewable resources and the development of technology has become a primary priority for the government of Algeria. The Algerian territory is characterized by its ample renewable energy sources, such as solar, biomass, geothermal, wind, and hydro. From this point of view, the Government of Algeria seeks to promote and accelerate the wider and more diversified participation of the private energy sector in the growth of investment, acquisition of technology, and access to foreign markets and know-how. This new policy required reforms of the government's legal and institutional frameworks, both at the global and sectoral level, to resolve the barriers facing efforts to boost the use of renewable energy sources, which need to be understood and properly translated into a coherent regulatory framework. Through this research paper, during the 2015–2019 period, we will try to define environmental indicators

S. Ahlem (✉)
LDEBG Laboratory Guelma University, University of 8 Mai 1945-Guelma, Guelma, Algeria
e-mail: soudani.ahlem@univ-guelma.dz

A. Zina
LAPDEC Laboratory Mascara University, University Ahmed Zabana, Relizane, Algeria
e-mail: zina.arabeche@univ-relizane.dz

M. El Amine Abdelli and M. Shahbaz (eds.), *Digital Economy, Energy and Sustainability*, Green Energy and Technology, https://doi.org/10.1007/978-3-031-22382-2_5

in the Algerian context, despite the considerable efforts made by the Government of Algeria to tackle the challenges of reducing economic dependency on hydrocarbon sales and greenhouse gas emissions. This paper offers a valuable framework for the status of environmental indicators in Algeria, which is considered a scientific forum for industrial companies and researchers to look for suitable solutions to promote environmental indicators. We have the following problem through the highlights mentioned above: What are the reality and prospects of environmental indicators in Algeria? The purpose of this study is to identify principles relevant to measures of the climate and know the truth of Algeria in terms of environmental and renewable energy indicators. Showing the significance of the paper through the significance of the issues discussed by:

- The relevance of the environmental indicators issue.
- The renewable energy sector's value as the future energy.
- The value in Algeria of the issue of environmental indicators.

The work approach: The preparation of this paper was based on the descriptive and analytical methodology in which, in the description of research variables, we find the descriptive form, the analytical method applied in the case study.

This research is organized as follows: Part 1 will identify the environmental performance definition metrics. Part 3 will address the reality of environmental performance in Algeria, followed by a brief discussion and the conclusion of this paper in Part 4.

2 Environmental Performance and Environmental Performance Evaluation Concept

Environmental performance is characterized as a multidimensional term that is difficult to measure as it relates to the environmental proposal to reduce the adverse effects of greenhouse gas emissions (Dkhili 2019). In addition, emissions, life-cycle evaluations, third-party audits, or environmental performance index (EPI) are considered, including both single factor indices and total factor indices. Environmental performance is characterized as a measurable outcome for managing the environmental aspects of an organization in compliance with the ISO 14001: 2004 standard. Results may be measured against the environmental targets and other environmental performance parameters of the organization (Peng 2019). The new ISO14031 standard defines environmental performance assessment (EPA) as an internal process, and management tool designed to provide reliable and verifiable information to be continuously controlled to determine whether the environmental performance of an organization meets the criteria set by the management of the organization (Christine 2000). In fact, the relationship between business and the environment can be characterized as measuring exchange and interaction (Giovanni 1996). To help define its environmental aspects, identify the aspects that are most relevant to it, set standards for its environmental performance, and assess its environmental performance based on these parameters, an entity that does not have an EMS can rely on the EPA to help define its environmental aspects.

The environmental performance assessment (EPA) is an important method for reviewing, auditing, tracking, and diagnosing the environmental performance of entities based on an environmental management framework. It is a continuous process of collecting and analyzing data and information to determine performance and patterns over time. The EPA is, therefore, a method of promoting management decisions on environmental issues, which relies primarily on the selection, compilation, review, diagnosis, and assessment of indicators in conjunction with the environmental performance measure and the preparation of the required reports to give us a complete and accurate environmental performance assessment (Mark 2000).

3 Indicators of Environmental Performance: Concept and Classification

3.1 Indicators of Environmental Performance Concept

The researchers differed in the definition of environmental performance indicators, which resulted in many indicators, as the concept of environmental performance that was adopted is multidimensional in nature, and this research was in specific areas of activity and depends on their own technologies, and therefore, the resulting indicators address problems linked to their specificity industry. The indicators reflect the degree of the organization's commitment to the protection or ruin of natural resources, which enables us to assess the extent of environmental damages related with the factories of the company and the degree of alignment with the institution's green strategy (Giovanni 1996). They have also defined it as a basic term that offers information on the environmental performance of an entity (Eleonara 2008). Environmental metrics are focused on the use of concrete performance measures. They provide valuable and standardized environmental performance data for evaluation comparison with various other site conditions (Le and Tam 2007). They have also been defined as a method for assessing system behavior in terms of both intended and perceived features. An indicator is, therefore, a measurement tool that collects and summarizes information about a specific variable or problem (Eleonara 2008).

3.2 Characteristics of Environmental Indicators

Environmental indicators, like other indicators, have a set of characteristics as follows:

They should be objective, important, meaningful, and involve all related issues; meet stakeholders' needs; help conduct meaningful comparative studies. Also, they should be applicable, and the information we need for the application can be obtained easily(Giovanni 1996).

3.3 Classifications of Environmental Performance Indicators

Environmental performance indicators are divided into three categories according to the ISO 14031 standard: operational performance indicators (OPIS), management indicators (MPIS), and environmental condition indicators (ECI), to provide local, regional, national, or global information (Mark 2000). (Loew and kottmann, 1996) classify environmental indicators (EIs); the metrics used at the microeconomic level to define environmental performance evaluations can be subdivided into financial and non-financial indicators (Ogrodnik 2019):

- The organization relies on a series of environmental and financial metrics to measure the profitability of green activities. Indicators based on the net value of future flows, internal rate of return (IRR), cost and benefit analysis, cost efficiency.
- Non-financial indicators define the interrelationships between enterprises and the environment; so, according to ISO 14031, non-financial indicators are divided into:
  - Management Performance Indicators (MPI): productivity indicators relating to the compatibility of provisions relating to social relationship management activities and services.
  - Operational Performance Indicators (OPI): system input indicators (materials, energy, and operational activity support facilities, technical devices, and equipment) and system output indicators (SOIs) (products, waste, emissions, services provided by organizations),
  - Health indicators for the environment: air, water, soil, biodiversity, and flora.

The OPIs were divided into four key parts of the company's basic resource inputs and outputs: material use, energy use, unproductive outputs, and emissions of pollutants. This scheme could cover the characteristic features of companies from various industries. We also hypothesize that EOP is measured by two OPIs: inputs and outputs. Inputs refer to the products used or consumed by a business (e.g., water, raw materials, paper, packaging materials, and hazardous chemicals) and energy (e.g., oil, electricity, gas). The more the organization uses resources/energy, the more waste/pollutants it can generate. In addition, the quality of using resources/energy in EPE should be considered to allow the company to increase its production performance. Outputs cover pollutant pollution produced by the company's operations, such as the disposal of general and industrial waste, the discharge of water, the discharge and transport of hazardous chemicals, the release of airborne pollutants (carbon dioxide, sulfur oxide, and nitrogen oxides), and the contamination of water

(COD or BOD) (Shuangyu and Kohji 2007). It is important to develop an adequate collection of indicators to properly analyze and evaluate the effect of environmental aspects; so in this study, we try to evaluate environmental performance in Algeria through its most important indicators, which are CO_2 emissions, energy production, and consumption, electricity generation, renewable energy.

4 The Reality of Environmental Performance in Algeria

4.1 Carbon Dioxide (CO_2) Emissions

The emissions of carbon dioxide are those arising from the burning of fossil fuels and cement manufacturing. They include carbon dioxide emitted by solid, liquid, and gas fuels and gas flaring during consumption. In estimating CO_2 emissions, only because of the difficulty of quantifying certain variables such as speed, weather, circumstances, driving style, congestion, and road types, CO_2 computation can reflect an approximation (Andres 2020). Under the Paris Climate Agreement adopted in 2015, Algeria is one of the most important CO_2 emitters among developing countries and the third among African countries to reduce carbon emissions by at least 7% by 2030 (Mohammed and Maria del Pablo 2016) (Fig. 1).

Algeria is a major oil and gas producer and exporter; the volume of oil exports accounts for a very large percentage of total exports, as the average percentage of its contribution during the period (2001–2017) accounted for about 95% of total revenues, while non-hydrocarbon exports account for just 2% of total exports (According to the annual reports of the central bank of Algeria).

Algeria has witnessed a development in carbon dioxide emissions in recent years for the following reasons (Fig. 2):

- The development of manufacturing industries and industrial construction;
- The development of the transport sector: roads, railways, and cross-country transport; an increasing number of cars and public and private transport; an increasing

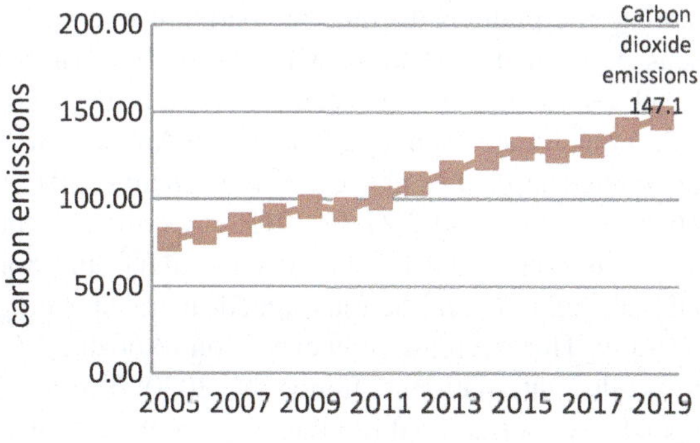

Fig. 1 Algeria carbon dioxide emissions (2005–2019). *Source* BP Statistical Review of World Energy June 2020

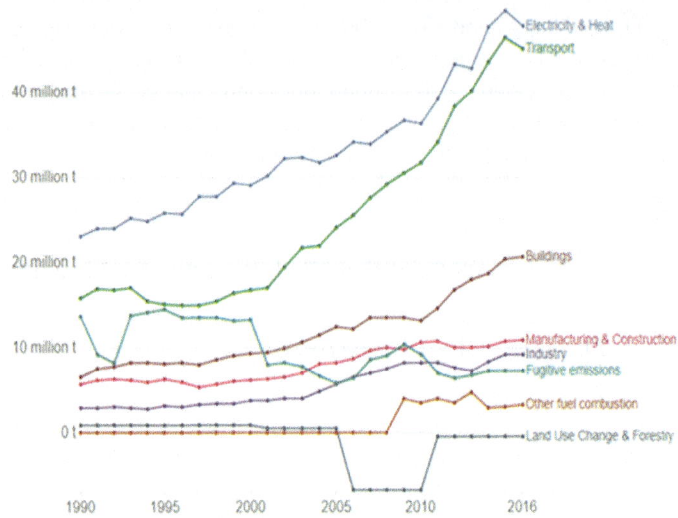

Fig. 2 Fossil CO_2 Emission by sector. https://ourworldindata.org/grapher/co-emissions-by-sector?time=earliest..latest&country=~DZA

number of domestic and international airlines, local navigation, random fishing … etc.;

- The number of residential buildings and commercial activities has increased in recent years as a result of the policy adopted by the state to eliminate fragile housing and access to decent housing for every member of society, with the emergence of new reconstruction policies specific to remote areas;
- Increased liquid fuel consumption from the use of oil-derived fuels as a fuel source;
- Increased volume of fugitive emissions from vapors, compressed gases from the hydrocarbon sector, and mining, processing, storage, and transportation of coal in large enterprises.

In Jordan, Algeria has reiterated its intention to lower its greenhouse gas emissions during 2020–2030 in line with the 2015 Paris Climate Agreement, which aims to stabilize the planet's average temperature rise at 2 °C. The Algerian ClimaSouth focus point at the Algerian Ministry of Foreign Affairs, which is taking part in the 6th conference of the executive board of the "ClimaSouth" program (European project on climate change) conducted in Amman on 29–30 March, remembered that Algeria had applied its national commitment in 2015 to decrease its GHG emissions between about 7 and 22% in 2015. According to the Algerian report, from 2020, Algeria is committed to reducing its GHG emissions by 7% through the use of national methods and actions aimed at transforming energy and diversifying the economy. This decrease, however, would touch 22% so that Algeria would receive the needed international support by 2030. The accomplishment of this objective depends on international funding, technical assistance, and capacity building as part

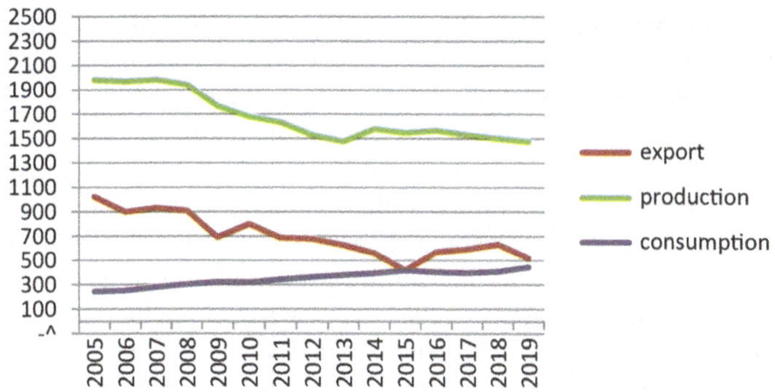

Fig. 3 Hydrocarbon production, export, and consumption (2005–2019). *Source* BP Statistical Review of World Energy June 2020

of the Algerian report. Algerian contribution includes the three major pollution gases, namely carbon dioxide (CO_2), methane (CH_4), and nitrous oxide (N_2O). Algeria is betting on a cohesive policy targeting the electricity, forestry, infrastructure, transport, industry, and waste sectors to achieve the reduction targets.

4.2 *Hydrocarbon Production, Export, and Consumption*

Algeria relies heavily on revenues derived from the petroleum industry, accounting for approximately one-third of global state revenue in 2016 (Latest International Monetary Fund Statistics).

Oil production (in thousand barrels per day) increased from 1347 in 1990 to 1992 in 2007; its highest value is attributed to the increase in wells discovered in 2007 as a result of the state's efforts, in addition to hydrocarbon insurance law 07/05, which motivated and privileged foreign investors in the hydrocarbons field. After 2007, the global economic crisis of 2008–2009 impacted oil production, and there were consecutive falls from 2007 to 2019, with 1486 thousand barrels per day (Fig. 3). The dependency of the Algerian country on oil products for economic growth increases its vulnerability to crude oil price fluctuations. By diversifying its exports (hydrocarbon products accounted for around 93% of global exports in 2019)[1] and limiting its growth in energy consumption to 3% in 2030, the state economic policy for 2016–2019 limits its exposure to the volatility of crude oil prices (International Monetary Fund 2018).

Algeria's government maintained an estimated 12.2 billion barrels (b) of proven oil reserves at the beginning of 2018. As offshore exploration has already been restricted, all of the country's proven oil reserves are kept onshore. We note the

[1] Trade Statistics Outside Algeria,2019, https://www.douane.gov.dz/IMG/pdf/rapport_com_ext_2019_vf.pdf. p. 14.

decrease in oil output beginning in 2018 from Fig. 3. In 2016, Algeria agreed to limit its capacity to 1.04 million b/d as part of the OPEC production reduction agreement and was prepared to comply with this commitment. Algeria will not be affected by the decision at the OPEC conference in Vienna in June 2018 to allow countries that have been over-compliant to increase production levels, as it has no spare capacity. Although Sonatrach spent $2.3 billion at the end of 2018 on the problem of energy decline since 2014, to boost crude oil and gas output (Houcini 2019), increase well efficiency, and build IOR/EOR techniques in existing fields in the SH30 strategy, this major effort will be more effective in the case of efficient production processes.[2]

According to EIA estimates, Algeria exported about 528,000 b/d of petroleum products in 2019, based on the navigation systems data provided by Clipperdata. In 2017, the main export markets by volume were Europe and Eurasia, which received 57% of Algeria's export value, followed by Asia–Pacific and Western Hemisphere regions, which received 21% of Algeria's exports, respectively. By volume, importing 82,000 b/d and 92,000 b/d, respectively, the United Kingdom and France were the main export destination nations. America was one of Algeria's largest markets for petroleum products for almost a decade until 2013. Imports of United States petroleum products from Algeria have declined considerably in recent years. In 2017, about 66000 b/d of oil was imported from Algeria by the United States, down from a peak of 443,000 b/d in 2007. The growth of bright, sweet petroleum production from the shale plays of Bakken and Eagle Ford in the United States has caused a major decrease in United States imports of petroleum products from Algeria.

4.3 *Natural Gas Production, Consumption, and Exports in Algeria (2005–2019)*

An analysis of Algeria's natural gas balance in the last three decades reveals an uneven production profile. After the rapid growth of the mid-1990s, which resulted from political changes and the failure of policies in the late 1980s and early 1990s, consumer demand declined and then stagnated since the mid-2000s. It should be noted that the early downturn in the production of natural gas is consistent with the promulgation in 2005 of the most recent and controversial Hydrocarbons Act and the subsequent inauspicious reform in 2006 (Fig. 4).

Commercialized gas production decreased by approximately 11% between 2005 and 2013, from 85 Bcm to about 79 Bcm. It then stagnated for the next two years at 80 Bcm. We noticed that the production in 2018 (94 Bcm) showed a considerable increase compared to 2015 and 2016 (BP Statistical Review of World Energy June 2020). This was partly due to the back to the regular capacity of the gas processing at Tiguentourine (In Amenas), which was destroyed during the terrorist attack in 2013, and to other minor increases in the supply of gas, including from In Salah (The oxford institute for energy studies, 2019, p08). Algeria's declining trend in the production

[2] http://www.us-algeria.org/attachments/article/280/SH2030.pdf.

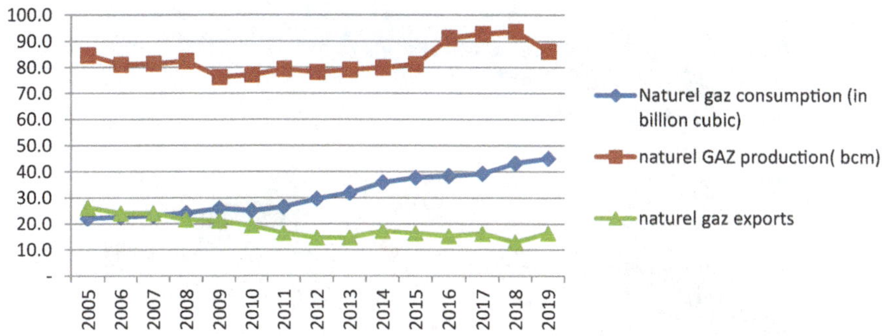

Fig. 4 Algeria's gas production, consumption, and exports: 2005–2019. *Source* BP Statistical Review of World Energy June 2020

of natural gas was not surprising. This demand-side deterioration has been predicted by specialists within and outside Algeria (Aissaoui 2016). A variety of reasons led to this decline:

- The old fields in natural gas production have become too mature, including the largest field Hassi Rmel. Also, some of the fields in the Berkine Basin that began to develop during the year 1990 are beyond their plateau production level.
- Lack of or inadequate investments to boost new lower recovery rates in secondary and tertiary recovery technologies.
- Low rate of new discoveries and evidence of new gas resources due primarily to an unfavorable environment in upstream investments for international investors.
- Bureaucratic concerns result in long gaps in the production and implementation of projects.

Algeria's natural gas output has recently been lifted regarding the construction of new reservoirs in the region's South West province. New products from this region cover three classes of areas that have recently come online and are projected to generate a total of nine Bcm a year at the plateau level, as detailed in Table 1

Table 1 Algeria's new south west gas fields

Project	Commissioning	Projected production (Bcm pa)	Consortium
Reggane Nord	December 2017	2.7	Sonatrach, Repsol, DEA10, Edison
Timimoun	March 2018	1.8	Sonatrach, Total, Cepsa
Touat	September 2019	4.5	Sonatrach, Neptune Energy11
Total		9.0	

Source The oxford institute for energy studies, 2019, p. 08

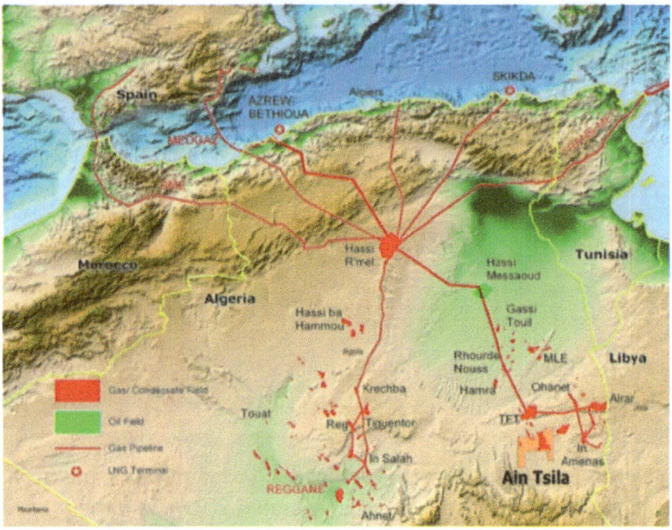

Map 1 Simplified natural gas map of Algeria. *Source* http://energy-cg.com/OPEC/Algeria/Algeria_Oil Gas_Industry.html

In the southwest zone, there are proposals to develop other new gas deposits (Hassi Mouina North and South, Ahner, Tidikelt South, and Akabli) (see Map 1). These are supposed to improve the supply availability from this region. Sonatrach company signed an EPC agreement in February 2019 with Larsen & Toubro Petroleum Engineering Limited of India for the installation of three main manufacturing plants (Oxford 2019).

Objectively needed additional supplies would be produced by the South West gas deposits, mainly to boost the gas supply to Algeria and then to contribute to upholding current domestic and export agreements. It is predicted that these gas reserves will play a significant role in new developments in the gas supply in Algeria. However, the cost of production in this new gas province is estimated to be higher than the supply from existing gas-producing fields in the South East region. This is due to the low volume of these fields, the limited availability, or the complete lack of infrastructure in this new, remote hydrocarbon field. Moreover, the gas produced is less rich in useful liquids containing natural gas and, in some instances, may have a high CO_2 content (The oxford institute for energy studies, 2019).

The deterioration of the natural gas balance in Algeria is motivated by the decline or stagnation of gas production and the rapid increase in domestic gas consumption. Total gas use in the domestic market grew by 70% between 2008 and 2018 at an average annual growth rate of over 5%. The public distribution market has increasingly increased the gas usage segment, as the "gasification program" remains a key government policy priority worldwide. The national gas penetration rate is currently above 60%, with the penetration rate close to 100% for the northern coastal population centers. Industrial consumption of gas consists of the sale of gas to two classes of industrial consumers (ibidem).

4.4 Renewable Energy in Algeria

Renewable energy is characterized as inexhaustible energy because it comes from sustainable sources such as the sun, water, wind, and oceans. Renewable energy is a clean alternative for building, heating, and cooling electricity generation that prevents air pollution reduces atmospheric carbon dioxide and greenhouse gas emissions. Usual or persistent natural phenomena caused by stars, such as the sun (radiation), as well as the moon (tides), and the Earth (geothermal energy), are the primary sources of renewable energy (EMER 2019) (Fig. 5).

Their common feature is that, in the exploitation process, they do not produce polluting (or small) emissions and therefore help to combat the greenhouse effect and global warming (EMER 2019). Alternative sources of electricity, including natural winds, solar, and geothermal energy, are cheap and healthy and have no harmful emissions. In addition, their preponderance is separate; there is no need to convey energy; the method is typically followed by major losses and losses over long distances, pollution of the atmosphere, like high-voltage electromagnetic pollution, and transport. Considering the high- and low-energy consumption, the introduction

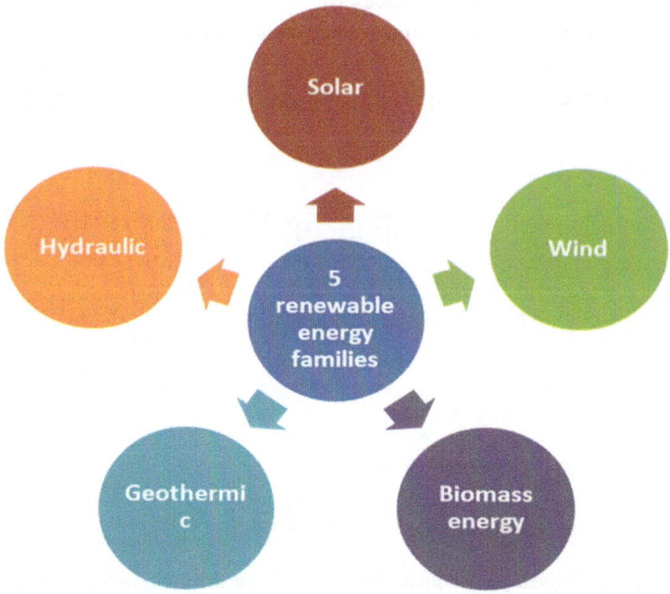

Fig. 5 Different types of renewable energy

of alternative energy sources is especially important for natural gas reserves (Kucher and Prokopchuk 2018), so the introduction of alternative energy sources is particularly important in Algeria. Algeria launched 2011 a national program to develop renewable energy based on photovoltaics (PV), concentrated solar power (CSP), and wind power to promote energy efficiency. To meet domestic electricity demand by 2030, the program requires installing up to 12 GW of power-generating capacity from renewable sources (Wikipedia 2014).

4.4.1 Solar Energy

With about 170 TWh per year, Algeria has the highest technological and economic potential for solar power exploitation in the MENA region. The inauguration of the Hassi R'Mel power station in 2011 launched the first industrial-scale solar thermal power plant. Combined with a 130 MW combined cycle gas turbine facility, this new hybrid power plant integrates a 25 megawatt (MW) concentrated solar power array (Wikipedia 2014) (Fig. 6).

Lower oil prices are already significantly affecting the implementation of renewable energy technologies in the power sector, even though policymakers fail to have the requisite market laws, policies, and subsidies. Indeed, the cumulative installed photovoltaic (PV) power (megawatts) increased from 25 in 2010 to 448 in 2019, owing to the decrease in oil prices from 96.29 in 2014 to 448 in 2019 (Saliha 2018) (Table 2).

Algeria's national program has set a goal of covering 40% of its national electricity needs from renewable energy resources by 2030. Due to its broad solar exposure, which covers 90 percent of the country with an area of 2382 million km^2 (Bouacha 2020), it has committed most of its renewable energy sources from solar photovoltaic PVS, and the desert occupies the forefront in producing the energy obtained,

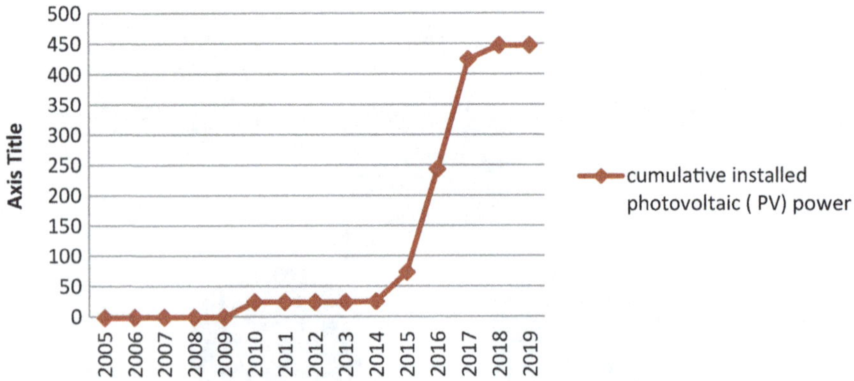

Fig. 6 Cumulative installed photovoltaic (PV) power: over the period (2005–2019). *Source* BP Statistical Review of World Energy June 2020

Table 2 Largest Algerian photovoltaic power stations

PV power station	Capacity MW	Notes
High Plateaus East, Adrar	90	Completed 2016, built by Sinohydro Corp (PowerChina), Yingli Green Energy Holding, HydroChina, owned by SKTM (Sonelgaz)
High Plateaus Center, Adrar	90	Completed 2016, built by Sinohydro Corp (PowerChina), Yingli Green Energy Holding, HydroChina, owned by SKTM (Sonelgaz)

Source Energy in Algeria Wikipedia, solar energy explained (Wikipedia Energy in Algeria 7, available at: https://en.wikipedia.org/wiki/Energy_in_Algeria (accessed 22 November 2020)

Table 3 Solar potential in Algeria

	Areas coastal	Area high plains	Sahara
Surface (%)	4	10	86
Area (km^2)	95.27	238.174	2.048.297
Mean daily sunshine duration (h)	7.26	8.22	9.59
The average duration of sunshine (h/year)	2650	3000	3500
Received average energy (kWh/m^2/year)	1700	1900	2650
The solar daily energy density (kWh/m^2)	4.66	5.21	7.26

Source Bouraiou et al. (2020)

estimated at 2650-kWh/m^2/year, as a result of having the highest brightness rate estimated at 3500 h per year, followed by the high plateaus region at a brightening rate of 3000 h per year, then the coastal region with 2650 h per year. As we find the solar daily energy density reaching up to 7.26 (kWh/m^2) in the Sahara (Table 3).

4.4.2 Global Horizontal Irradiation (GHI) in Algeria

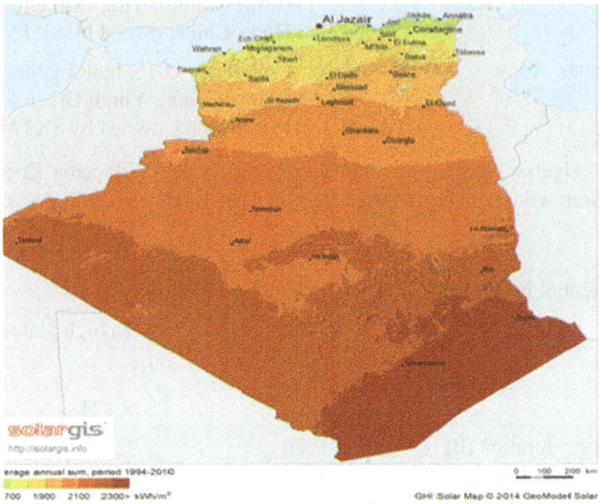

Global Horizontal Irradiation (GHI) in Algeria

Source Energy in Algeria Wikipedia (2014), solar energy.

4.4.3 Hydraulic Energy

It is the water flow energy that can be used to move the water wheels connected to generators to produce electricity; energy comes from water from many sources such as waterfalls and dams, wave energy in the oceans, as the action of waves produces energy that can be used and converted into electrical energy. The total estimated capacity of installed power is almost 270 MW, as detailed in Table 4.

4.4.4 Hydroelectricity Generation

Hydroelectricity dropped from 0.6 terawatt-hours in 2005 to 0.2 terawatt-hours in 2019. Apart from the droughts that Algeria has experienced in recent years, this poor capacity is attributed to the inadequate number of irrigation sites and the lack of optimum use of existing irrigation sites (Fig. 7).

4.4.5 Wind Power

Wind power is the energy produced by moving giant fans mounted on high-place columns by the movement of the air to provide mechanical power by wind turbines

Table 4 Hydropower installed in Algeria

Darguina	Ighil Emda	Mansouria	Erraguene	Souk El Djemaa	Tizi Meden	Ighzernchebel	Ghrib	Gouriet	Bouhanifia	Oued Fodda	Tessala	Total
71.5	24	100	16	8.085		4.458	2.712	7.000	6.425 5.700 15.600	3.500	4.228	269.20

Source Bouraiou et al. (2020)

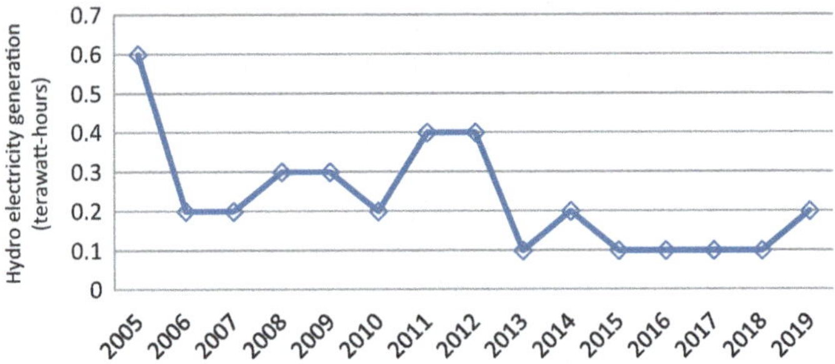

Fig. 7 Algeria hydroelectricity generation (2005–2019). *Source* BP Statistical Review of World Energy June 2020

to transform electric generators into electrical power. As the wind moves on the fans' blades, it causes them to rotate, and this rotation drives the turbines and creates electrical energy. In contrast to burning fossil fuels, wind power is a common sustainable, renewable energy source with a much smaller effect on the environment. In 2019, the global wind power market grew 19% to 60 GW, the second largest annual growth, for a total of 650 GW (621 GW onshore and the rest offshore) (report, 2020).

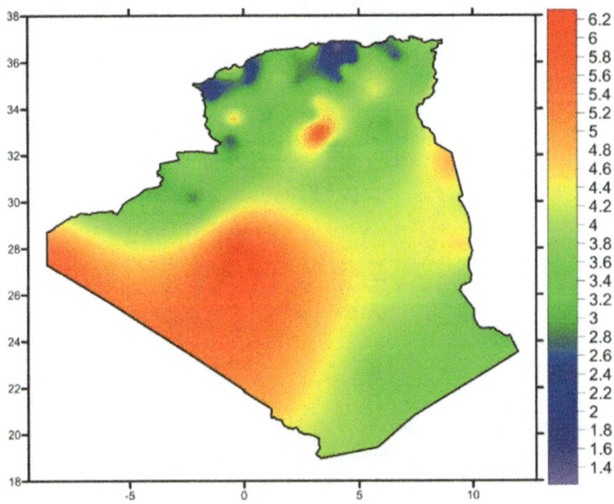

The wind fields in algeria

Source Ministry of Higher Education & Scientific Research General Directorate for Scientific Research and Technological Development, Renewable Energy Development Center.

4.4.6 Geothermic Energy

The compilation of geological, geochemical, and geophysical data has allowed more than two hundred (200) hot springs in the northern part of the country to be found. Around a third of them have temperatures greater than 45 °C (33%). In Biskra, there are sources of high temperatures exceeding 118°. Three regions with a gradient above 5 °C/100 m were defined in the thermal gradient studies: the Relizane region, the Ain Busif and the Sidi Isa area, area of Guelma and Mount Onek (Ministère de l'énergie 2020). To display more than 240 locations, the inventory of the thermal springs has been updated. The temperature of the hot waters of Algeria ranges from 22 to 98.8 °C. 68.8 °C for the western area (Hammam Bouhnifia), 808 °C for the central area (Hammam El Biban), and 98.8 °C for the eastern area (Hammam Meskhoutine) in northern Algeria which are the highest spring temperatures recorded. There are some thermal springs in the southern region with a mean temperature of 50.8 °C. In northern Algeria, the total dissolved solids (TDS) of the hot springs are greater than 1 g/L (see Map 4). The largest geothermal reservoirs in northern Algeria are carbonate formations, while the reservoirs are mainly composed of sandstone in southern Algeria. Three geothermal regions have been delineated based on the distribution of thermal springs and geological and geophysical considerations (such as permeability and geothermal gradient).

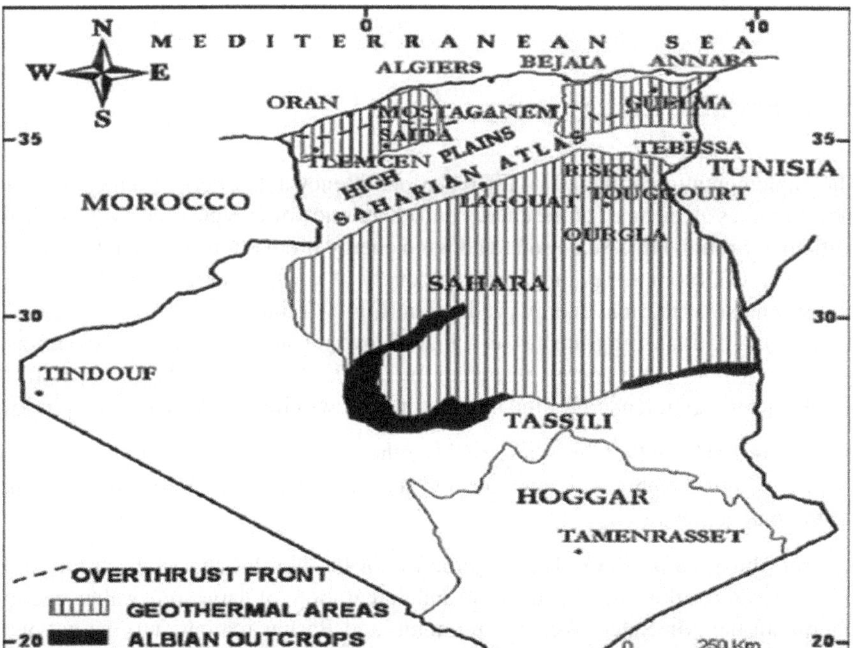

The location of the three geothermal regions of Algeria *Source* (Saibi 2009)

4.4.7 Biomass Energy

The oldest and simplest way of using energy is burning biomass. In terms of climate, biomass electricity and energy production are a particularly attractive method of conversion of energy. With 3.7 million TOE coming from forests and 1.33 million TOE coming from urban and agricultural waste annually (365 kg per Algerian as urban waste), biomass in Algeria offers excellent promises technically, but this potential is not yet enhanced and used. A pre-survey demonstrates the likelihood of 2 MW modal power generation that can reach a peak of 6 MW from the Algiers Oued Smar discharge. The analysis incorporates the location's drainage. MEM regulations encouraging the use of biomass from energy crops have rapidly led to an increase in interest in linking agriculture and energy sectors. This is seen as a first step in stimulating the use of biomass quite quickly in Algeria. Biogas is also considered a desirable and relatively inexpensive fuel source. In addition, the disposal of biogas by combustion is important for protecting the atmosphere, particularly to protect the environment from the emission of unbranded methane contained in biogas. The UREERMS (Solar Equipment Experimentation Unit in the Sahara Area) has started a gradual increase in the use of biogas, particularly from landfills, which is seen as a step forward in the use of biogas in Algeria. The gasification method is a very beneficial choice for burning. Using biogenic gas fuels, proven and efficient techniques such as gas turbines and cogeneration units can be applied (Abda and Bouharkat 2018).

5 Renewable Energy Policy

The implementation of the 2011–2030 National Renewable Energy Policy positions these energies at the heart of the energy and economic policies pursued by Algeria, in particular through large-scale solar and wind energy production and the introduction of branches of biomass (evaluating the recovery of waste), thermal and geothermal energy, and solar thermal energy production (Fig. 8; Table 5).

Algeria's national program has set a goal of covering 40% of its national electricity needs from renewable energy resources by 2030. Projects for the production of renewable energy intended for the national market will be introduced in two phases:

Phase one 2015–2020: This phase will see the achievement of 4010 megawatts of electricity between photovoltaics and wind and 515 megawatts between biomass and geothermal cogeneration.

Second phase 2021–2030: The construction of the North-Saharan electricity link (Adrar) would allow the installation and integration of large renewable energy stations in the Ain Saleh, Adrar, Timimoun, and Bashar regions into the national energy system. Solar thermal energy could be more economically feasible by this time.

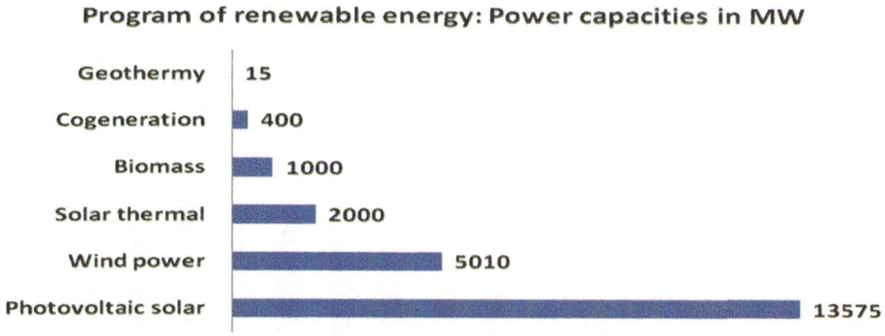

Fig. 8 Distribution of renewable energy by sector of energy production. *Source* Webinar series: sustainable energy in humanitarian settings, https://energypedia.info/wiki/Algeria_Energy_Situation#Energy_Data

Table 5 Algerian RE target

Source	1st phase 2015–2020 (MW)	2nd phase 2021–2030 (Mw)	Total (MW)
Solar PV	3000	10,575	13,575
Wind	1010	4000	5010
CSP	–	2000	2000
Cogeneration	150	250	400
Biomass	360	640	1000
Geothermal	5	10	15
Total	4525	17,475	22,000

Source Webinar series: sustainable energy in humanitarian settings, https://energypedia.info/wiki/Algeria_Energy_Situation#Energy_Data

Renewable energy capabilities will be installed according to the specifics of each region (Industry 2017):

- **The southern region**, to hybridize the existing centers and feed dispersed sites according to the availability of spaces and the importance of solar and wind energy capabilities.
- **Highlands region**, according to its capabilities of sunlight and wind, with the possibility of acquiring plots of land.
- **Coastal areas**, according to the possibility of the availability of real estate vessels with the exploitation of all spaces such as roofs, balconies, buildings, and other unused areas.

A national research program in this field has been set up to accompany the renewable energies development strategy, as the scientific objectives of this program aim at evaluating renewable energy deposits, controlling the process of transferring and

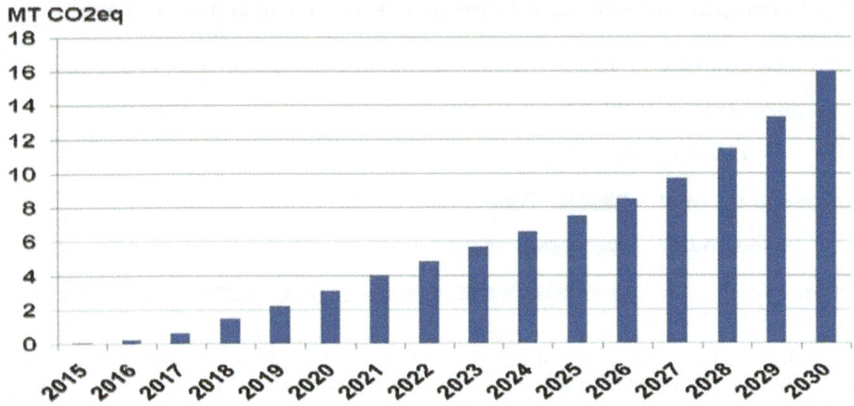

Fig. 9 Annual evolution of CO_2 avoided through the implementation of the renewable energy program. *Source* Sahnoune et al. (2017)

storing these energies, and developing the necessary skills, starting from the study until the completion of the completion at the site of installation (Fig. 9).

The total amount of CO_2 avoided is about 102 MT CO_2eq for the entire 2015–2030 period. This refers to the emissions from national energy production and use, considering all industries, for 2015. The CO_2 savings were also about four times the emissions from traditional power generation in 2015.

6 Result and Discussion

Renewable energy is seen in Algeria as a crucial driver of development that can help to address two of the nation's key challenges: economic diversification and job creation (Arabeche & Abdelli). Algeria has valuable energy reserves, mainly solar energy and crude oil. Following the assessment carried out by the German Aerospace Center, considerable ability appears to be available for using renewable energy sources, particularly solar and wind resources. However, the degree of growth of such energy power is very primitive, but efforts can expand due to the ever-increasing problem of sustainable development. To increase the use of this renewable energy, either in the public or private markets, and to illustrate the necessity of this clean energy, it is now important to inform society as much as possible to introduce and sustain special legislation on energy.

Algerian authorities must take into account the obligation to switch to renewable energy for the following reasons:

1. Algeria plays a capital role in fossil energies as much as a producer and exporter, even though it has potential in renewable energies.
2. Lack of a global, efficient, and stable trajectory for renewable energy programs.

3. As the national renewable energy potential is strongly dominated by solar energy, Algeria must consider this energy as an opportunity and a lever for economic and social development (Era 2020).
4. Improving conditions for private investment in renewable energy in different sectors in Algeria.

7 Conclusion

In recent years, the Algerian authorities have made a lot of efforts to mobilize the introduction of renewable energy through several initiatives (SKTM, GEG, CDER, etc.) while implementing a legal and regulatory structure to face several challenges, such as the fall in oil prices on international markets, global warming and the economy's reliance on hydrocarbons. The Algerian government plans to produce at least 40% of electricity from renewable energy sources by 2030. The generation of electricity from renewable energy sources is a real investment opportunity where a regulatory contract guarantees economic viability. Also, Algeria has one of the world's highest solar capacities due to its geographic location. The Hassi R'mel solar project presents NEAL's expertise par excellence in energy production and many national and private company-funded initiatives, all of which give Algeria a key role in adopting renewable energy technologies in Africa in particular and in the world in general. Among the recommendations concluded by the study are:

- Adopting a holistic approach to renewable energy with clear and observable targets and metrics, with an emphasis on green employment, technical advancement, R&D, incorporation of production, and improvement of natural capital;
- Improving the technical capabilities of organizations and organizations for renewable energy;
- To raise awareness in the media, schools, and universities of the value of renewable energy;
- The current economic scene in Algeria is marked by economic austerity, falling oil prices, social, and economic problems, a wide geographical region, and a large renewable energy opportunity, all of which invite decision-makers to reconsider their policies;
- Establishing a more profitable national innovation framework based on policies supporting innovation, especially for small and medium-sized enterprises with limited financial funds and skills;
- Supporting the transfer of technology in the context of collaboration (North–South and South–South).

Appendix

See Tables 6, 7, 8, 9 and 10.

Table 6 Carbon dioxide emissions (million tons of carbon dioxide)

	2005	2006	2007	2008	2009	2010	2011	2012	2013	2014	2015	2016	2017	2018	2019
Carbon dioxide emissions	76.9	80.8	85,2	90.6	95.7	94.2	100.6	108.9	115.4	123.6	129	127.7	130.7	140.4	147.1

Source BP Statistical review of world energy June 2020

Table 7 Energy production (2005–2019)

	2005	2006	2007	2008	2009	2010	2011	2012	2013	2014	2015	2016	2017	2018	2019
Oil production (in thousand barrels per day)	1990	1979	1992	1951	1775	1689	1642	1537	1485	1589	1558	1577	1540	1511	1486
Natural gas production (in billion cubics)	84 0.9	81.3	81.6	82.6	76.6	77.4	79.6	78.4	79.3	80.2	81.4	91.4	93	93	86.2

Source BP Statistical review of world energy June 2020

Table 8 Energy consumption (2005–2019)

	2005	2006	2007	2008	2009	2010	2011	2012	2013	2014	2015	2016	2017	2018	2019
Oil consumption (in thousand barrels per day)	249	258	286	309	327	327	349	370	387	401	425	412	407	417	454
Natural gas consumption (in billion cubics)	22.4	22.8	81,6	24.4	26.2	25.3	26.8	29.9	32.1	36.1	37.9	38.6	39.5	43.4	45.2

BP Statistical Review of World Energy June 2020

Table 9 Electricity generation over the period (1990–2019) (trawatt-hours)

	2005	2006	2007	2008	2009	2010	2011	2012	2013	2014	2015	2016	2017	2018	2019
Electricity generation trawatt-hours	33.9	35.2	37.3	40.2	43.1	45.7	53.1	57.4	59.9	64.2	68.8	71	76	76.7	81.3

Source BP Statistical review of world energy June 2020

Table 10 Renewble energy solar: cumulative installed photovoltaic (PV) power (megawatts)

	2005	2006	2007	2008	2009	2010	2011	2012	2013	2014	2015	2016	2017	2018	2019
Cumulative installed photovoltaic (PV) power	–	–	–	–	–	25	25	25	25	26	74	244	425	448	448

Source BP Statistical review of world energy June 2020

References

Abda Z, Bouharkat M (2018) Study pf management strategy of energy resources in Algeria. Energy Rep 4:1–7
Aissaoui A (2016) Algerian gas: troubling trends, troubled policies. Oxford Institute for Energy Studies, NG 108
Andres M (2020) Measuring environmental performance of urban fright transport systems: a case study. Sustain Cities Soc 52:3
Arabeche Z, Abdelli M (s.d.) Insight of green economy in Algeria. Dans freindly energy processes and technologies for achieving sustainable development. IGI Global, pp 108–126
Bouraiou A et al (2020) status of renewable energy potential and utilization in Algeria. J Clean Prod 06:1–16. https://doi.org/10.1016/j.jclepro.2019.119011
Bouacha S (2020) Performance analysis of the first photovoltaic grid connected system in Algeria. Elseiver, p 57
Christine J (2000) Environmental performance evaluation and indicators. Journal of Cleaner Production 79
Dkhili HD (2019) Management of environmental performance and impact of the carbon dioxide emissions (CO_2) on the economic growth in the GCC countries. Marketing and Management of Innovations
Eleonara P (2008) Environmental performance, indicatores and measurment uncertainly in EMS context: a case study. Journal of Cleaner Production 518
EMER (2019) Ministry of environment and renewable energies. National Institute of Environmental Compositions. Consulté le 12 27, 2020
Era (2020) Programme de développement des énergies renouvelables
Giovanni A (1996) Defining environmental performance indicators: an integrated framework. Bus Strateg Environ 5:69–80
Houcini A (2019) Oil and gas mining domain of Algeria activities: opportunities and assessment of potential. Algeria Energy Days. Houston
Industry M (2017) National agency of investment development, renewable energy sector. Consulté Le 12(14):2020
International Monetary Fund (2018) Algeria charts new economic direction. is it battling against the tide? vol 60, Issue 18. Middle East Economic Survey. 5 May 2017
Kucher OO, Prokopchuk L (2018) The development of the markets of the renewable energy in Ukraine. In: Proceedings in Energy. Springer international Publishing AG
Le KN, Tam VWY (2007) Assessing environmental performance in the construction industry. Surveying and Built Environment 18(2):59–72
Leow T, kottmann H (1996) kennzahlen im umweltmanagement. okologisches witschaften 2. https://scholar.google.com/scholar_lookup?&title=Kennzahlen%20im%20Umweltmanagement&journal=%C3%96kologisches%20Wirtschaften&volume=2&pages=10-12&publication_year=1996&author=Loew%2CT&author=Kottmann%2CH#d=gs_cit&t=1672047652754&u=%2Fscholar%3Fq%3Dinfo%3APBMRSWm4LLUJ%3Ascholar.google.com%2F%26output%3Dcite%26scirp%3D0%26hl%3Dfr
Mark O (2000) ISO 14031 effective mechanism to environmental performance evaluation. Corporate Environmental Stategy 7:272
Ministère de l'énergie (2020) Récupéré sur Energie nouvelles renouvelables et Maitrise de l'energie
Mohammed B, Maria del Pablo R (2016) CO_2 emission and economix growth in Algeria. Energy Policy 93
Ogrodnik R (2019) Environmental performance indicators of hard coal mine. In: 2nd international conference on the sustainble energy and environmental development, IOP, conference series: earth and environmental science, p 03
Oxford (2019) Oil & gas business. Consulté le 12 26, 2020, sur Oil & Gas Business
Peng TB (2019) Impact of financing constraints on firm's environmental performance: evidence from China with survey data. J Clean Prod 217:433

Report G (2020) Renewables. Ren 21
Sahnoune F et al (2017) Energy consumption renewable energy development and environmental impact in Algeria—trend for 2030. In: AIP conference proceedings, vol 1814. American Institute of Physics, p 020072. https://doi.org/10.1063/1.4976291
Saliha H (2018) Algerian energy policy: perspectives, barriers and missed opportunities, global challenges, vol 2, issue8. Willey VCH Verlag GMBH and KGaA Weinheim, p 1
Shuangyu X, Kohji H (2007) Corporate environmental performance: a measurement model and a new concept. Business Startegy and the Environment, p 156
Wikipedia (2014) Energy in Algeria: solar energy. Consulté le 11 22, 2020, sur Wikipedia
Wikipedia (2014) Energy in Algeria: wikipedia. https://fr.wikipedia.org/wiki/%C3%89nergie_en_Alg%C3%A9rie

Circular Economy in Algeria: Strategies and Obstacles

Adel Fatima Zohra and Guendouz Abdelkader

1 Introduction

The actual tendencies toward the circular economy aim to realize a set of achievements favoring sustainable development. The various schools of thought confirmed the goals of the circular economy increasingly appeared in the late twentieth century, thus highlighting the importance of the circular economy as a new economic model (Braungart 2002). The circular economy depends on the change in the patterns of production and consumption in a more rational way, which seeks to preserve valuable resources, materials, and products for the longest possible period. Circular production and consumption occur by closing loops and dealing differently with waste (Walter and Stahel 1981), where it is not disposed of as in a linear economy, but rather it is reintegrated into production chains as a material that has been recycled or reused in consumption after it has been repaired (Ellen Macarthur 2012). On their path toward sustainable development principle commitment, most countries seek to shift to this model, which requires them to adopt the strategies that form the basis of this transformation. However, this process seems to be more complicated for developing countries, as in the case of Algeria. Two decades ago, Algeria made great strides in the pursuit of sustainability by taking a set of measures aimed at integrating the environmental dimension as a basic component of its economy, as well as eliminating various obstacles facing this integration, thus contributing to building a strong ground for the transition into a circular economy. In this paper, we will present the various circular economic strategies and enumerate the obstacles and challenges facing Algeria, which must be invoked to establish the rules of sustainability that pave the way toward the transition into a circular economy in future. Suppose the

A. F. Zohra (✉) · G. Abdelkader
Tahri Mohamed University, Béchar, Algeria
e-mail: adel.fatimazohra@univ-bechar.dz

G. Abdelkader
e-mail: guendouz.abdelkader@univ-bechar.dz

M. El Amine Abdelli and M. Shahbaz (eds.), *Digital Economy, Energy and Sustainability*, Green Energy and Technology, https://doi.org/10.1007/978-3-031-22382-2_6

transformation to a circular economy is a strategic choice under the commitment to sustainable development principles. In that case, this transformation requires a change in production and consumption patterns. Governments must adopt a set of policies and strategies that support this change. As for Algeria, its transformation to a circular economy requires mastering the environmental dimension and waste management, which poses major challenges that must be lifted.

Our study aims to know what obstacles prevent Algeria from adopting strategies to transition to a circular economy and how they can be lifted. This chapter is based on an analytical approach consisting of basic requirements determination circular economy strategies, then determining whether they are susceptible to implementation or not for the Algerian case.

2 Literature Review

Geissdoerfer et al. (2017) argued that while the terms circular economy and sustainability are gaining increasing traction with academia, industry, and policymakers, the similarities and differences between both concepts remain ambiguous. The relationship between the concepts is not made explicit in literature, blurring their conceptual contours, and constraining the efficacy of using the approaches in research and practice. This research addresses this gap and aims to provide conceptual clarity by distinguishing the terms and synthesizing the different types of relationships between them. We conducted an extensive literature review, employing bibliometric analysis, and snowballing techniques to investigate state of the art in the field and synthesize the similarities, differences, and relationships between both terms. We identified eight different relationship types in the literature and illustrated both concepts' most evident similarities and differences. Bey et al. (2013) noted that for environmental strategies to take effect in industrial practice, they must be implemented and applied in daily business procedures. Based on a comprehensive international survey dedicated to product development and manufacturing companies, this research identified the main current drivers for implementing environmental strategies oriented toward the product lifecycle and the barriers and obstacles that must be addressed. On this basis, it makes several recommendations for manufacturing companies and policymakers to consider the successful implementation of strategic environmental objectives in the manufacturing industry. Masi et al. (2018) defined the circular economy and proposed an economic framwork based on circular flows of products and materials. Understanding the nature and status of this transition is important for creating effective policies and action strategies. This exploratory, survey-based study of 77 companies investigated the shift toward the CE using a taxonomy of practices and barriers. Results showed that firms favor resource and energy utilization efficiency, while practices related to investment recovery, green purchasing, and customer cooperation are less prevalent. Eco-design and internal environmental management practices have a medium level of implementation. The significant upfront investment cost, lack of awareness, or urgency were identified as implementation barriers. Results suggested economic, not environmental, considerations drive the CE, and the deployment of practices remain within a firm rather than across the supply chain.

3 Circular Economy

Many firms and international institutions have used the circular economy concept as a new alternative economic model for the linear one; each defined the circular economy term. However, Korhonen (2018) defines the circular economy as a sustainable development initiative to reduce the societal production-consumption systems' linear material and energy throughput flows by applying materials cycles and renewable and cascade-type energy flows to the linear system. CE promotes high-value material cycles alongside more traditional recycling and develops systems approaches to the cooperation of producers, consumers, and other societal actors in sustainable development work. Plus, the Ellen MacArthur Foundation defines an industrial system that is restorative or regenerative by intention and design. It replaces the "end-of-life" concept with restoration, shifts toward the use of renewable energy, eliminates the use of toxic chemicals, which impair reuse, and aims to eliminate waste through the superior design of materials, products, systems, and, within this, business models. The objective is to enable effective flows of materials, energy, labor, and information to rebuild natural and social capital.

Moreover, European Environmental Agency defines the circular economy as providing opportunities to create well-being, growth, and jobs while reducing environmental pressures. The concept can be applied to all kinds of natural resources, including biotic and abiotic materials, water, and land (European Environment Agency 2016). The European Committee in a circular economy and the value of products and materials is maintained for as long as possible. Waste and resource use are minimized, and when a product reaches the end of its life, it is used again to create further value. This can bring major economic benefits, contributing to innovation, growth, and job creation. Despite the diversity of definitions and approaches to the circular economy, it can be summed up in two main features. One is its opposition to the linear economic model, and the second is its belonging to the sustainable development framework. The circular economy is also characterized by relying on the concept of circularity in the extraction, production, use, and recycling of waste after consumption, which constitutes a closed circle. Through this circle, waste is reduced to the lowest level, using new methods: circular design, repair, restoration, and reuse of resources and energy. All this is within a sustainable framework that preserves the environment without stopping economic growth.

4 Circular Economy Strategies

The circular economy is based on the idea of change in both consumption and production patterns.

It aims to preserve the value of resources, materials, and products by reducing waste for the longest possible period. The circular economy includes three fields of intervention: The first concerns the production and supply of goods and services where the industry is based on the sustainable procurement of resources—whether renewable or fossil—and depends on the ecological design of products, which allows it to contribute to reducing waste at the industrial and regional environment level. It also focuses on the functional economy. The second field of intervention relates to all sides of consumption, responsible demand, behavior, purchasing processes, participatory consumption, sound use, restoration, and repair.

The third field of intervention involves waste management, which is inevitable to consolidate recycling and energy valorization, even if responsible use and production greatly contribute to reducing waste production (reduction at source). The implementation of a circular economy in a society or country must be done by adopting strategies (Hagelüken et al. 2016).

4.1 Functional and Sharing Economy

If the circular economy seeks to change production and consumption patterns, the functional and the sharing economy form the basis for this change. Each of them includes a variety of commercial strategies and exchange models, allowing a steady increase in the use of traded products in the market. While the sharing economy relies on sharing or redistributing products among consumers, a functional economy suggests that the firm keeps its ownership of the products but leases their use. Functional economy refers to the process of focusing on the function of products to make them used by a large number of consumers; this was found before the circular economy in many industries for long periods, such as libraries and car agencies. Currently, through a functional economy, firms keep ownership of the asset and sell the service instead because lease contracts replace the transfer of ownership. Many firms started to adopt this model, such as Xerox, which suggests selling copies instead of printers, and Michelin, which proposes to sell kilometers instead of tires (Stahel 2010). The sharing economy aims to promote the sharing and re-exchange of goods, and although this system is as old as economic, social, and commercial transactions, nowadays, it takes other more expanded and organized forms due to the spread of technology and digital platforms and the move toward the environmental dimension. This expansion is more observed in parallel markets, where the cooperative economy stems from the self-regulatory behaviors of consumers or associations (Stahel 2016).

4.2 Maintenance, Repair, and Reuse

These concepts are interconnected, especially when it comes to a product. Within the framework of a circular economy, maintenance is intended to carry out periodic control of a product or equipment to maintain a good performance of its function, thus ensuring its durability. As for repair, it has focused on addressing a breakdown in a product or production equipment to reduce the throwing in a way that guarantees the recovery of products for reuse or recycling if they cannot be recovered in their original form (Bras and McIntosh 1999). The firms use the eco-design to facilitate the product components disassembling process after its consumption, then keeping the valid parts and recycling the damaged parts. Sorting at the source reduces the next stages of the waste treatment process to ensure a set of regulatory procedures related to hazardous waste.

4.3 Ecology and Industrial Symbiosis

The interest in industrial ecology results from the increasing pressure to eliminate negative environmental impacts and keep raw materials from depleting. Industrial ecology aims to reform industrial systems through innovative strategies that integrate ecology with technology. It constitutes a multidisciplinary framework for designing and operating industrial systems under the same bases and rules of natural systems and in an interconnected manner with them. Natural systems are considered ideal for managing natural resources, including waste and energy, making industrial systems part of natural ones (Frosch 1989). In addition to the decent appearance of the environment and the urban aspect, adopting an industrial system allows maximum wealth for the least consumption of raw materials. Industrial ecology can be considered a manufacturing policy of the best production, better exploitation of resources, and production waste so that this exploitation becomes more profitable and efficient (Erkman 1997). The new concept of industrial ecology—which means achieving the production process as a closed cycle—is the symbiosis of the industrial process in recycling materials in the optimal way, which seems to be the key element in the emergence of models of the relationship between the industrial environment and nature (Dewulf 2013). Because of its structure based on a series of industrial interconnections, such as energy exchange, raw materials, water treatment, waste recycling, and other industrial linkages within industrial cities, industrial symbiosis is regarded as one of the most important elements of industrial technology. Industrial symbiosis is organized in a manner that simulates natural systems, where traditionally separate entities participate and tangle in a collective approach to acquire a competitive advantage, which enables the reduction of waste and pollution and favorite the efficient sharing of resources such as information, energy, materials, water, infrastructure, and natural resources. One example of industrial symbioses is the ecosystem in Kalundborg (Denmark); in this ecosystem, the residual product of one company is used as resources for other companies in a closed circle.

4.4 *Recycling and Valuation of Waste*

The strategies of the circular economy also concern the after-waste creation phase. In the case that other strategies become unable to extend the life of the product (repair and maintenance, eco-design ...), it is necessary to recover the resources (materials and energy) present in the products at the end of their life cycle to reintegrate. These resources are in the value cycle of the circular economy (Geldron 2012). Product recycling strategies, including organic products recycling, allow their re-exploitation in the production of new product and their reuse.

5 Obstacles to Implementing Circular Economy in Algeria

The adoption of various strategies for the circular economy is not as simple as it appears. Many challenges prevent the implementation of these strategies, which differ according to the adopting entity, because institutions may face a set of internal obstacles on the one hand and external obstacles related to the environment in which they are active on the other hand. The responsibility for implementing the circular economy rests primarily in charge of the government due to its ability to formulate different policies with its power to create actors' commitment to implementing them.

Algeria is considered a developing country, where adopting a circular economy constitutes real bets that require overcoming obstacles to be lifted. The obstacles that face Algerian firms in the process of implementing a circular economy can be listed as follows:

5.1 *Technical Constraints*

The circular economy depends largely on the activity of managing and recycling waste, but this latter faces a set of technical obstacles in its functioning. Waste recycling allows for reducing the extraction of raw materials, but it cannot stop the current irrational exploitation of resources. The same observation can be valid for closed recycling, which is also limited, considering that the quality of the raw materials decreases after each recycling process. In Algeria, many waste management programs have been approved in addition to the enacted laws, such as the National Program for Domestic and Similar Waste Management 2001 "PROGDEM" and the National Plan for special Waste Management 2001 "PNAGDES". These programs are highly important regarding their goals and are used to push the field of waste management. The first program has allowed organizing, collecting, transporting, and disposing of solid waste, ensuring the protection and preservation of the environment, and developing and equipping underground waste backfilling center (Ministère de l'environnement and Agence Nationale des déchets 2020). The second program was management, planning, and supporting decision-making tools by providing quantitative and qualitative information on everything related to waste production and its development in Algeria (National waste agency 2020). Although these programs and laws exist and the multiple measures taken to control waste in Algeria to protect and preserve the environment, some deficiencies still impede the good functioning of the waste management process. The percentage of waste in 2016 reached about a 34millon tons, which is likely to increase, as the proportion of waste may rise from 0.8 kg to more than 01.23 kg per inhabitant per day by 2035 (Cheniti 2014).

The total subsidized annual cost of managing waste by the state reached about 73 billion Algerian dinars in 2016. Moreover, the resources allocated to this process are publicly funded through the annual domestic waste TAX TEOM and other taxes, even if they are insufficient to meet the public financial equilibrium. This insufficiency is due to their low percentages on the one hand and their weak collection, which does not exceed 15%, on the other hand, which constitutes a great burden on the state. This situation worsens in the absence of a private sector that faces organizational and administrative obstacles that hinder it from investing in this sector (Djemaci 2012).

5.2 Financial and Economic Obstacles

The functional and participatory economy is one of the most important circular economy strategies, as both aim to reduce throwing and increase product life through repair, reuse, and recycling. Consumers' perception of life-extended products is negative because their prices are close to the prices of the new products, and their performance is bad since they were used previously. In some ways, they are also considered outdated styles and unfitting fashion; in addition to qualifying recycled products as "dirty waste", the behavior of consumers is not the only matter, but their negative impact on economic growth can be problematic. This seems to be explained by the functional and participatory economy principle itself, which may lead to a reduction in sales and then a reduction in production, which negatively affects economic growth due to its direct correlation with the volume of new products and the added values resulting from production. The situation in Algeria regarding the functional and sharing economy is still traditional and far from being considered a circular economy strategy. This can be confirmed by the fact that consumers in Algeria are, till now, dealing with old goods through reuse, which may be caused by low purchasing power levels many times, but also somehow they think that old goods are of high quality in comparison with modern ones.

In addition, the repair principle is still carried out by retired craftsmen who have gained experience from their previous work in various public industrial sectors. As already mentioned, this process extends the life of products by repairing and reusing them. All of this cannot reach a level of circular economy strategy, at least in the short term.

5.3 *Social and Cultural Obstacles*

The circular economy is a modern concept that considers waste as a resource. This new conceptualization of waste remains difficult to be accepted by society in all of its components, such as industrialists, civil society, and businessmen because the term waste is usually a synonym for dirt. Therefore, it is not easy to adopt a circular economy and seek to implement its own strategies, especially when closing the chain value and integrating waste as a resource within this chain. Currently, in Algeria, the situation is positively changing toward reconsidering waste as a resource (National Waste Agency 2020) through the sensitization done by both the media and education and training institutions on the one hand, as well as encouraging all kinds of private initiatives in the field of dealing with waste on the other hand. It also depends on the media in order to encourage investment in these sectors, through programs that allow monitoring all the incentives adopted by the state, especially since the whole world is heading towards this activity, which has become profitable at the present time.

Despite all these activities aimed at changing the people's negative perception of waste, they are merely initiatives that have not the necessary professional tools to transform the collective awareness of Algerian society in its different classes and various educational, professional, and social levels toward a circular economy.

5.4 *Legislative and Regulatory Barriers*

Legislation is the most important tool by which the state intervenes to organize society, as it is, at the same time, a means used for deterrent and stimulus purposes. A large number of legislative texts regulate the environment-related sectors for safe use and exploitation, especially after adopting the "the polluter pays principle". From the beginning of the 1990s, the waste management problem in Algeria started to attract attention as a central public issue. The government apprehended the necessity to regulate this environmental problem, which imposed the adoption of many reforms in various related sectors using the legislative instrument. In 2001, executive decrees issued and reinforced the 01–19 law act related to waste management, control, and removal. The application of these decrees faced organizational and structural problems with ambiguity in defining the tasks and interference in authorities in the field of waste management and removal services. At an institutional level, the absence of the private sector exacerbates the situation in waste management. While governments in other countries have shifted the burden of taking the waste charge of the private sector, the government in Algeria is still the only achiever of the entire waste management process, whether at the financial, technical, or even legislative levels. In addition, the sectors that intersect in managing waste as a common responsibility have not a unified strategy for sharing this burden. Also, because of legislation, some authorities, such as municipalities, lack the power to intervene in waste management, as they do not have the right to collect financial allocations that allow them to set up and supervise programs at the local level. Since its creation, the Algerian Ministry of Environment has played an important role in environmental management. Many programs have been adopted under this ministry that has effectively and positively contributed to protecting the environment on the one hand and regulating waste management on the other hand. However, it still suffers from insufficiency due to the overlapping of its functions with other ministries, which provokes a dispersion in its activities.

5.5 The Taxes Problem

Taxes are one of the most important means that can play a crucial role in the circular economy since they are used both as a stimulus and deterrent tool. It pushes those undergoing it to change their production and consumption behaviors in favor of reducing pollution, but sometimes, it creates obstacles for investors in the field of the circular economy. If misused, in terms of the determination of tax rates, assets concerned with that tax, and exonerations, taxation can negatively affect the adoption of a circular economy in a society. Like other countries, Algeria has used tax to encourage sustainability through environmental taxation. The Algerian government adopted this latter through legal frameworks and devices installed to ensure its implementation and collection. An example of these devices is the National Environment Fund, which was established under the Finance act 1992, which collects revenues from taxes imposed on violators of environmental legislation.

In 1992, a set of deterrent taxes had been enacted in Algeria, such as a tax on polluting activities and those hazardous for the environment, as well as taxes on each household waste, new rubber tires (regardless of their original), fuels, etc. In parallel, other incentive taxes were issued in 2020; such as the tax encouraging destocking of health waste within hospitals and tax on private wastes generated by industrial firms that are dangerous to the environment for industrial establishments (Article 203 of the Finance Law in 2002), where the tax value is linked to the amount of waste (30,000DZD, and 60,000DZD per ton, respectively), according to art. 89 and art. 90 of the finance act 2020. Environmental taxation in Algeria faces some contradictions in the manner of imposing it, its proportions, determining the parties responsible for damaging the environment, and the difficulty of obliging those undergoing it to pay it. Despite all the taxes that have been issued, environmental taxation is still far from fulfilling its role in controlling waste, which constitutes an obstacle to the circular transformation economy.

6 Conclusion

The circular economy is considered an alternative new economic model to the prevailing linear economic system, through which the principles of sustainable development in production, consumption, and waste management are adopted in ways that allow waste reduction. It has the potential to fit economic dynamism with environmental difficulties. Circular economy strategies must be adopted to ensure success in the transformation to circularity, as these strategies allow a complete change in the patterns of production and consumption in a way that allows extending the life of products on the one hand and manages waste on the other hand, by reducing it from the source, reusing or recycling it to become new products. To answer the research problem related to identifying the most important obstacles that prevent the adoption of strategies for the transition to a circular economy, this paper concluded by accepting the basic hypothesis of the research: Indeed, many obstacles must be lifted for Algeria to move to a circular economy. These obstacles can be identified in the following points:

Technical constraints: The waste management process in Algeria, which requires a lot of money to control, is not covered by the resources allocated to it and the absence of private investment in this sector.

Economic and financial constraints: The participatory functional economy in Algeria is still far from being a circular economy strategy, as the consumption pattern is still traditional.

Socio-cultural obstacles are related to Algerian citizens' behavior toward waste and increasing product life cycle through reuse, repair, and maintenance.

- **Legislative and regulatory constraints:** The implementation of the various legislations issued by Algeria in the environment faces organizational and structural problems with ambiguity in defining tasks and overlapping authorities in the waste management and removal services.
- **The taxes issues:** Environmental taxation in Algeria faces some contradictions in the manner of imposing it, its proportions, determining the parties responsible for damaging the environment, and the difficulty of obliging those undergoing it to pay it. Implementing the circular economy in a society is not as simple as it appears. Many obstacles and challenges must be lifted before adopting the circular economy. Being a developing country, Algeria cannot create a circular economy dynamic, at least in the short term, despite the presence of some indicators that can form a basis for adopting this new economy. The Algerian involvement in various agreements that support sustainable development is among the most important of these indicators, as well as the adoption of the "polluter-payer principle" announced in 1972 by the Organization for, Cooperation and Economic Development to lower emissions and reduce waste. Algeria's transition to a circular economy requires it to prepare the basis on which this economic model can be established and gradually eliminate the obstacles that prevent this transformation,

then adopt the necessary strategies to realize the circular economy through a given interest in developing the important elements of each strategy, and identifying the various tools that are used in each of them. Controlling the waste management process is urgently necessary to facilitate the transition to the circular economy. This objective can be achieved according to the following recommendations:

- Allocating infrastructures for waste management and introducing institutional change by redividing tasks and determining responsibilities for all entities in which this process intersects, as well as strengthening and updating various laws and procedures that would further regulate the waste management sector and an amendment to the law on waste.
- Since the circular economy depends on managing and recycling waste, the technical means and methods necessary for this process must be developed, especially by encouraging local investment in this sector and adopting modern concepts related to waste problems, such as the green economy and cleaner production.
- Working on providing various mechanisms that push firms to reduce their waste and recycle it in deterrent or stimulating ways.
- Working on promoting the role of municipalities in dealing with household waste and involving various actors in this process to contribute to better management of local waste.
- Reconsidering environmental taxation, especially when it comes to imposing and collecting the tax. The value of TEOM, for example, is very low, and its rates must be increased, as well as to strengthen its collection methods.
- Developing solidarity and sharing economy can qualify as stagnation in a traditional stage in Algeria. This situation is to be taken as background susceptible to forming a starting point of the circular economy through transforming solidarity and sharing practices from amateur actions to more professional programs.

References

Agency, National Waste (2020) Waste is resource,waste management magazine for Africa and Mena region. Algiers, Algeria

Bras B, McIntosh MW (1999) Product, process, and organizational design for remanufacture—an overview of research. Robotics and Computer-Integrated Manufacturing 15(3):167–178

Braungart WM (2002) Cradle to cradle: remaking the way we make things. North Point Press, New York

Cheniti H (2014) la gestion des déchets urbains solides: Cas de la ville d'Annaba. Annaba,Algérie: these de doctorat

Dewulf K (2013) Sustainable product innovation: the importance of the front-end stage in the innovation process. In: Coelho PD (ed) Advances in industrial design

Djemaci B (2012) La gestion des déchets municipaux en Algérie : Analyse prospective et éléments d'efficacité, Thèse de doctorat. Université de Rouen, France

Ellen Macarthur F (2012) Towards the circular econom, vol 3: accelerating the scale-up across global

Erkman S (1997) Industrial ecology: an historical view. Journal of Cleaner Production

European Environment Agency (2016) Circular economy in Europe—developing the knowledge base; EEA, Copenhagen, Denmark
Frosch RA (1989) Strategies for manufacturing. Sci Am 261:144–152
Geissdoerfer M, Savaget P, Bocken NM, Hultink EJ (2017) The circular economy–A new sustainability paradigm? J cleaner prod 143:757–768
Geldron A (2012) Peut-on recycler à l'infini ? Pour la Science
Hagelüken C, Lee-Shin JU, Carpentier A, Heron C (2016) The EU circular economy and its relevance to metal recycling. Recycling 1(2):242–253
Korhonen JN (2018) Circular economy as an essentially contested concept. J Clean Prod 175:544–552
Masi D, Kumar V, Garza-Reyes JA, Godsell J (2018) Towards a more circular economy: exploring the awareness, practices, and barriers from a focal firm perspective. Prod Plann Control 29:6:539–550
Ministère de l'environnement et l'agence nationale des déchets (2020) Bulletin de veille technologique N°03, Alger, Algérie
Stahel W (2010) The performance economy. Palgrave-MacMillan, London
Stahel WR (2016) The Circular Economy. Nature News 531(7595):435
Walter R, Stahel GRM (1981) Jobs for tomorrow: the potential for substituting man power for energy. Vantage Press

Development and Energy

Commercializing Bee Pollination to Increase Maize Productivity and Farmers' Economic Gains in Tanzania

Proscovia Paschal Kamugisha, Kubwela Mwangu Rumulika, and Robert John Mwenyasi

1 Introduction

Maize plays numerous roles in human lives as it provides nutrients for humans and animals and serves as the primary raw material for producing various products such as alcoholic drinks, oil, protein, fuel, starch, and food sweeteners (Wu and Guclu 2013; Zibaee 2013). Major nutrient contents of maize include starch (72%), protein (10%), and fat (4%). The rest (14%) contains various vitamins and micronutrients, including vitamin B and other essential minerals, providing 365 kcal/100 g (Nuss and Tanumihardjo 2010).

Globally, maize is one of the most important and widely cultivated crops as it plays a great role in meeting food demand; that provides about 50% of consumed calories (Macauley and Ramadjita 2015). Maize plays dynamic roles in food and nutrition security and agri-food systems across the globe (Grote et al. 2021). Worldwide, about 56% of maize produced is used for livestock feeds, 20% for non-food uses, and 13% is directly consumed as human food (Erenstein et al. 2022). While the use of maize in the global north is mainly for livestock feeds, in Sub-Saharan Africa and Tanzania in particular, maize is a staple food which constitutes more than 40% of the typical meal components (Cochrane and D'Souza 2015). Different relevance of maize in food and/or nutrition security in different parts of the world is reflected in its annual per capita consumption, which amounts to 64 kg in Eastern and Southern Africa compared to 35.9 kg in Central and Southern America, 7.9 kg in Europe, and 10.1 kg

P. P. Kamugisha (✉) · K. M. Rumulika · R. J. Mwenyasi
Department of Business Studies, Mbeya Campus College, Mzumbe University, Mbeya, Tanzania
e-mail: ppkamugisha@mzumbe.ac.tz

K. M. Rumulika
e-mail: kurumulika@mzumbe.ac.tz

R. J. Mwenyasi
e-mail: rmwenyasi@mzumbe.ac.tz

M. El Amine Abdelli and M. Shahbaz (eds.), *Digital Economy, Energy and Sustainability*, Green Energy and Technology, https://doi.org/10.1007/978-3-031-22382-2_7

in Asia. Tanzania's annual maize per capita consumption is estimated at 112.5 kg (Temu et al. 2011).

In Tanzania, maize is grown by 4.5 million smallholder farmers holding farm sizes ranging between 0.9 ha–3 ha (Kumari 2018) in more than two third of the country's arable land, which accounts for over 90% of the national total maize production (Nyaligwa et al. 2022). Maize production accounts for 50% of cash to Tanzanian rural dwellers, which amounted to USD 100 for maize-producing households in 2008 (Barreiro Hurle 2012). However, its production depends entirely on rainfed since only three percent of the cultivated land area is irrigated (United Republic of Tanzania (URT) 2016). Coupled with a lack of practicing good agricultural practices, Tanzania's maize productivity ended up ranging between 1.2 t/ha and 1.6 t/ha (Nyaligwa et al. 2022), which is extremely low compared to the potential average maize productivity of 7.5 t/ha (Magehema et al. 2015; Kihara et al. 2020) and below the global average maize yield of 3.9 t/ha as well as that of Africa (85.5 t/ha) (Epule et al. 2022; Erenstein et al. 2022). Furthermore, the Tanzanian low maize productivity is projected to decline by 37% in 2050 if the business as usual continues (Arndt et al. 2012; Msongaleli et al. 2015) due to the changing climatic conditions. The current and projected low maize productivity is expected to put more pressure on non-expanding country agricultural land to feed the ever-growing national population.

It is projected that, national population will increase from 61,280,743 in 2022 to 86,871,546 in 2035 (United Republic of Tanzania (URT) 2018) and 135,244,161 in 2050 (Mwakisisile and Mushi 2019). The increased national population translates national food requirements to increase from 8.5 million tons in 2022 to 12 million tons in 2035 and 18.7 million tons of grain equivalency in 2050. This is based on the minimum food requirement of 138 kg of grains equivalency/annum (Beghin 2015). Actually, Lyimo (2005) projected an increase of 3.5% in maize demand per annum to meet national food requirements in the next 20 years.

The World Bank's (2018) estimation of cereal productivity of 1544 kg/ha reflects the expansion of agricultural land area from 5,505,181 ha in 2022 to 12,111,398 ha in 2050 to feed the national residents at the agricultural land expansion rate of 235,936 ha per annum. Since the national land is limited, increased agricultural land expansion can lead to serious land use conflicts between crops and livestock farmers (Benjaminsen et al. 2012; Falanta and Bengesi 2018; Rweyemamu 2019).

In developing countries like Tanzania, crop productivity is improved through fertilizers, pesticides, and irrigation, excluding pollination (URT 2022). However, the consensus among agricultural scientists views agricultural intensification by using inorganic fertilizers and pesticides as an insufficient strategy which can lead to environmental concerns in future (The Montpellier Panel 2013). Hence, it is argued that sustainable agricultural intensification entails production of more crops in the same land area with minimal environmental effects (Charles et al. 2014; Garibaldi et al. 2019). Bee pollination is thought to increase agricultural production while maintaining the environment. This book chapter advocates commercialization of bee pollination to improve maize productivity, increase farmers' income in Tanzania, and increase food security. The chapter's proposition is anchored within Tanzania's Intended Nationally Determined Contribution (INDC) (URT 2015) to the Paris Climate Agreement for post-2020 mitigation and adaptation action. The agreement promotes increasing agricultural yields through climate-smart agriculture (URT 2015) to attain current human needs without compromising the needs of future generations.

Studies on managed honey bee pollination in Tanzania are hardly documented. Several authors aligned their studies to beekeeping activities for honey production (Giliba et al. 2020; Godfrey 2018; Tutuba et al. 2019), maize production technologies and food security (Kassie et al. 2014; Magrini and Vigani 2016; Nkonya et al. 1998), and handling of post-harvest losses of maize produced (Abass et al. 2014; Suleiman and Rosentrater 2015). The studies regarding improvement of agricultural yields using honey bees pollination are mainly clustered in the horticultural, pulses, and oil seeds sub-sectors (Elisante et al. 2020; Sawe et al. 2020; Tibesigwa et al. 2019) excluding maize. However, the studies conducted employed wild honey bee pollination whose habitats are degraded as the population increase. To the authors' knowledge, no study in Tanzania has evaluated the role of managed honey bee pollination on maize productivity. Hence, this book chapter complements the present body of knowledge by shed lighting investment potentials to commercialize managed honey bee pollination services for increasing maize productivity to meet food and nutrition security requirements and improve smallholder farmers welfare.

2 Literature Review

The challenge of low productivity for staple food, including maize, in Tanzania can be traced back to the 1960s when the nation faced severe food insecurity. Different stakeholders: Government, researchers, non-governmental organizations, and other development partners devised several strategies to address low maize productivity and tackle food insecurity. These initiatives include formulation of the regulatory framework to underpin development of maize, developing different maize cultivars that suit more Tanzanian environment; and devising & implementing a number of strategies & programs to improve maize yield.

In this regard, the government of Tanzania has devised policies, strategies, and programs to amend the situation. The policy and legal agenda include the Agricultural and Livestock Policy (1997) and the National Agricultural Policy of 2013. Other Acts are the Fertilizer Act of 2009, the Plant Protection Act of 1997, and the Plant Protection Regulation of 1998. Moreover, several strategies and programs were implemented to raise agricultural productivity, including maize yields. The programs and strategies include: "*Kilimo cha Kufa na Kupona*," which means literary agriculture as a matter of life and death during villagization in the mid-1970s, National Maize Research Program-1974, Agricultural Sector Development Programme I (2006/07–2013/14), National Agricultural Inputs Voucher Scheme (NAIVS) of 2009, Big Results Now (2013) and Agricultural Sector Development Program II (2017/18–2025/26). All these policies, Acts and programs were geared to increase agricultural farm productivity and commercialize agriculture. Recently, the government has launched its National Beekeeping Policy Implementation Strategy 2021–2031, which intends to increase crop productivity through managed honey bee pollination.

Likewise, efforts to improve maize yield through research and development started during the 1960s. It was during this time when Ukiriguru Composite A (UCA), Ilonga Composite White (ICW) and improved Open Pollinated Varieties (OPV) were released (Katinila et al. 1998). In the early 1970s, the nation faced severe food shortages due to drought and farmers' displacement campaigns, "villagization" (Bagachwa and Naho 1994). To mitigate food shortages, the National Maize Research Programme (NMRP) aimed to develop maize cultivars that suit major maize production areas according to their agroecological zones. From the NMRP, several maize cultivars were released, including Tuxpeno, TMV2, Hybrids H6302 and H614, suitable for highlands; Staha was recommended for midaltitude, while Kito and TMV1 was meant for both midlatitudes and lowlands. To date, several maize cultivars are in place being produced not only by National Research Institutes but also by private companies that were born during the mid 1990s after market liberalization. Tanzania's existing private seed companies include PANNAR, SeedCo, Suba Agro Monsanto Seeds, and Highland Seeds that command 28%, 26%, 9%, 7%, 6%, and 6% of the maize seeds market share, respectively (Beyene and Kassie 2015). Generally, 63 hybrids and 21 open-pollinated varieties were developed and released since independence, 31 of which were developed by NMRP (Luzi-Kihupi et al. 2015). The released maize seed cultivars were more resistant to diseases and could adapt to changing climatic conditions.

Some of the programs and strategies made to address crops productivity, including maize, are summarized in Table 1.

It is acknowledged that development of various maize cultivars and adoption of other technological development such as the use of inorganic fertilizers & farm yard manure, application of pesticides & herbicides have increased maize yield (Frederick et al. 2020; Kihara et al. 2020). Despite the efforts above geared to improve maize yield, to date, productivity of maize farms is still far below (1.5 tons/ha) (Nyaligwa et al. 2022) compared to its production potential of 7.5 tons/ha (Magehema et al. 2015; Kihara et al. 2020). Hence, in addition to the efforts above, this chapter advocates the use of managed honey bee pollination in maize farms to improve maize productivity in Tanzania.

Table 1 Programs addressing maize productivity in Tanzania

Period	Government Initiative	Description	Actual yield	Author
1966 to the 1980s	Arusha Declaration (1967) Guidelines	Encouraged socialism and a centralized economy. Displaced farmers in "*Ujamaa villages*" to provide social services and subsidized improved inputs to increase agricultural productivity	Decreased productivity from 20 bags/acre (3.7 t/ha) in 1950s to 6 bags/acre (0.9 t/ha) in 1970s	Temu and Ashimogo (1999), Baregu (1987), Keskin and Abdalla (2019)
1990 to 1995	Input Markets Liberalization	Some farmers started to adopt improved maize seeds. This was expected to increase maize productivity to range between 3 tons/ha and 4.5 tons/ha	The productivity increased, yet not to the expected outcomes. The yield realized was less than 1.5 t/ha	Katinila et al. (1998)
2001 to 2005	Agricultural Sector Development Strategy (ASDS)	Identified required specific interventions such as improving water supply management and irrigation, farm mechanization, use of improved seeds, inorganic fertilizers and agrochemicals to increase agricultural growth rate by 5% per annum	Low productivity of about 0.08 t/ha in 2007 was caused by low financing, limited inputs and poor technologies	URT (2001); Pauw and Thurlow (2011)

(continued)

Table 1 (continued)

Period	Government Initiative	Description	Actual yield	Author
2009/10 to 2012/13	National Agricultural Input Voucher System (NAIVS)	It was a policy measure to implement ASDP to improve agricultural productivity that would range between 3.5 tons/ha–9 tons/ha based on maize cultivar, the humidity of the area and agronomic practices	Slight increase in maize yield of 1.2 tons/ha. Increase the use of inorganic fertilizer by 8–10%	Kim et al. (2019), Hepelwa et al. (2013), Frederick et al. (2020)
2015/16 to 2024/25	Agricultural Sector Development Program phase II (ASDP II)	More improvement in agricultural productivity. For maize, it is projected to increase the yield ranging between 3.5 tons/ha and 9 tons/ha based on maize cultivar, humidity, and applied agronomic practices	Still, with an average maize yield of 1.8 t/ha under fertilizer treatment and 0.6 t/ha under no fertilization	URT (2015); Mkonda (2022); Frederick et al. (2020)

3 Policy, Legal, and Institutional Frameworks for Bee Pollination in Agricultural Production

3.1 Policies, Legal Framework and Other Initiatives Governing Bee Pollination

Globally, the Convention on Biological Diversity (CBD) of the early 1990s and the International Pollinator Initiative Plan of Action (IPIPA) (2018–2030) recognize pollination as an important tool to promote Intensive Sustainable Agriculture (ISA) (CBD 2018). The African Pollinator Initiatives (API), founded in 1999, were launched to facilitate participation of African countries in global initiatives on sustainable conservation; and the use of pollinators to increase agricultural production. Tanzania's regulatory framework reflects CBD and IPIPA by integrating her policies, Acts, rules, and guidelines governing crops and beekeeping industries. Moreover, these policies and legal frameworks vested powers upon governed institutions that normally undertake various strategies and programs.

The national policies and legal framework governing pollination services in agricultural production are housed mainly in the Ministry of Agriculture, Ministry of Lands, Housing and Human Settlement Development, and Ministry of Natural Resources and Tourism. While agricultural sector is widely governed by the National Agricultural Policy of 2013, beekeeping activities are mainly governed under the National Environmental Policy of 2021, which was preceded by the National Environmental Policy of 1997. The regulatory framework governing agricultural sector and environmental issues emphasizes improvement of agricultural production while conserving natural resources and biodiversity to meet current needs without compromising needs of future generations.

The National Environmental Policy of 2021 governs environmental issues, including pollinators under biodiversity conservation (URT 2021c; 2021). The policy is operating in line with the Environmental Management Act of 2004 (EMA). Sector ministry legislations support Environmental Management Act. of 2004. Although there is no specific item on pollination in the policy, pollinators' economic role is stipulated under sections 1.2.3 and 3.3 on wildlife habitats and biodiversity conservation (URT 2021; 2004). At this point in time, the policy recognizes the importance of wild life resources to lie behind their economic value and potential contribution to National Sustainable Development. Section 3.3 of the policy portrays the intention of the government to strengthen regimes of biodiversity protection and utilization for optimizing benefits from wildlife.

The EMA of 2004 was developed by then as a working tool to implement the National Environmental Policy of 1997. Although the nation has launched the new Environmental Policy 2021, the Act still governs the new policy. Part V of the EMA 2004 stipulates natural resource management and sustainable use of the resource. In its Caption 66–68, the Act specifies promotion of conservation and management of biological diversity, including pollinators.

Furthermore, the National Beekeeping Policy of 1998 recognizes animal pollination's role as a useful tool for increasing crop yields, as stipulated in section 4.3.4 of the policy (URT 1998). Moreover, the National Beekeeping implementation strategy of 2021–2031 promotes the use of pollination as an instrumental tool to increase farmland productivity. In section 4.4, the strategy aims to increase the use of pollination services from 168 farmers in 2021 to 500 farmers in 2031 (URT 2021a). Currently, the government has developed the guideline for honey bees' pollination in Tanzania in May 2022 to guide implementation of the strategy (URT 2022). Likewise, the National Agricultural Policy 2013 promotes sustainable use of endowed natural resource bases for agricultural development. In section 3.1.3 (iii), the policy promises to support initiatives geared to arrest those involved in agro-biodiversity deterioration (URT 2013).

Generally, it can be concluded that honey bees' pollination is within the national policies and legal framework. Hence, whoever wants to invest in beekeeping and agricultural pollination services is allowed.

3.2 *Institutional Framework*

The power to coordinate wildlife and biodiversity conservation in the country is vested in the Vice President's Office, Division of the environment (VPO-DoE), given the current policy, legal, and regulatory framework. The beekeeping activities are ratified under Forestry and Beekeeping Division housed under the Ministry of Natural Resources and Tourism. Key players in honey bee pollination services include the Ministry responsible for environment, the Ministry of Agriculture, the National Environment Management Council (NEMC), the Regional Secretariat, and Local Government Authorities (URT 2021b). Others are the private sector, development partners, and academic and research institutions. Ministry of Agriculture and other key players have mainstreamed the compliance of the policy in their rules, guidelines, strategies and programs. Meanwhile, NEMC is in charge of ensuring adherence to the environment's legal requirements and execution of Environmental Impact Assessments for various programs undertaken in the country.

4 The Current Maize Production Practices in Tanzania

Maize is grown in all agroecological zones of the country using various agricultural technologies. The maize production depends on rainfed using longer maize varieties in areas experiencing unimodal rains and short-term varieties in the bimodal zones whereby three percent of production is under irrigation national wide (Xiong and Tarnavsky 2020; Tarnavsky et al. 2018; URT 2016; Wilson et al. 2015). A peculiar feature exists in southern highlands whereby a substantial proportion of smallholder farmers (23%) in Songwe irrigates their maize farms (Bongole 2021).

Different cropping patterns in maize farming are practiced in Tanzania. A substantial proportion (81%) of maize farmers in Lake and Western Zone uses monocropping, while some (21%) employ intercropping of maize with other crops (Lobulu et al. 2019). Sawe et al. (2020) noted that farming households in northern Tanzania grew an average of four crops practicing a monocropping regime for maize farms in rented lands and intercropping with legume crops in the owned farms surrounding their homesteads. Wilson et al. (2015) contended that maize farms are integrated with pigeon peas in the Southern highlands. Alternatively, Mather et al. (2016) revealed that in Tanzania, more than one-third (37%) of maize smallholder farmers employed monocropping, while 35% intercropped maize with legumes. Moreover, Bongole (2021) discovered that 34% of maize farmers intercropped maize with legumes, while 66% employed crop rotation.

The low use of improved maize seed varieties is experienced in most parts of the country. The commonly grown seed cultivars include Yanga, Katumbili, Gembe, SeedCo, Stuka, Bunane, Kilima, Pannar, 4M-19, Ukibgam Ukiliguru, Nyati, Pundamilia, Lubango, Pionneer, DKC 8053 (Lobulu et al. 2019). The low use of improved maize varieties is associated with high costs attached to hybrid seeds and economic risks since the improved seed varieties necessitate proper agronomic practices for viable investment (Adam et al. 2020).

Employment of inorganic fertilizers is still below standards in the country. In central and western zones of Tanzania, almost half (49%) and 37% of maize farmers use inorganic fertilizers and farm yard manure, respectively (Lobulu et al. 2019; Mather et al. 2016), despite a huge number of ruminants raised in their area. Actually, less than 10% of farmers employs inorganic fertilizers (Xiong and Tarnavsky 2020). In the Lake Manyara basin, a substantial proportion of farmers (78%) employed inorganic fertilizers, while 43.4% utilized farm yard manure (Kikiwete et al. 2015). Moreover, Bongole (2021) noted the use of residue retention (46.5%) and organic manure (36%) as major means of fertilizing agricultural fields southern highlands of Tanzania.

5 Potentials of Increasing Maize Productivity Using Managed Bee Pollination

Pollination entails movement of pollen from male anthers to female stigma, either within the same flower (self-pollination) or between plants (cross-pollination) (Khalifa et al. 2021). Pollination plays a great role in agriculture and serves as a basic pillar for crop production. Naturally, plants depend on various pollinators such as water, wind, insect pollinators, and honey bees. Insect pollination is the most efficient in increasing productivity worldwide compared to other pollinators. Gallai et al. (2009) discovered that the economic values of insects' crop pollination account for 10% of total agricultural production values globally. In Europe, the United States, Indonesia, and South Africa, the contribution of wild honey bee pollination and managed bee pollination amounts to US$ 3251/ha and US$ 2913/ha, respectively (Kleijn et al. 2015). Moreover, in Brazil, Giannini et al. (2015) found that bee pollination values of agricultural products account for 30% of agricultural incomes, amounting to US$ 45 billion annually among pollinators-dependent crops.

In Tanzania, Sawe et al. (2020) hold the economic importance of wild pollinators for horticultural crops in Arusha gardens. The authors noted that the gardens were pollinated by wild honey bees (87.8%), hoven flies (8.5%), and other Hymenoptera (3.7%). Despite its importance in pollinating crops, management of bees in the region is limited to the development of empty hives for honey collection. Actually, Sawe et al. (2020) revealed economic benefits of honey bee pollination for agricultural farms, including increased farm productivity. Yet, about two third of farmers grew crops close to forestry by chance without noting the benefits of wild bees. Elisante et al. (2020) have noted increased yield for beans farms from 681 kg/ha for farms without pollination to 1478 kg/ha for pollinated farms in Northern Tanzania. Meanwhile, the practice of crop pollination is done in other crops believed to depend mainly on animal pollination (Elisante et al. 2020; Giannini et al. 2015; Sawe et al. 2020) such as fruits and vegetables, oil seed crops, and some perennials excluding maize.

Although Tanzania is endowed with numerous wild bee colonies and a number of bees keeping activities, these bees are not rented out to improve agricultural productivity (Tibesigwa et al. 2019). Therefore, it is proposed that managed honey bee pollination services be commercialized to increase maize productivity (URT 2022) for sustainable access of the service among smallholder farmers. It is hypothesized that commercialization of managed honey bee pollination services would serve dual purposes: (1) Raising maize farm productivity as pollination is considered the incremental input in farming activities and (2) increase of honey harvests to beekeepers. The increased amount of honey harvested is thought to be because, in the course of pollination, honey bees will be securing their feeds (nectar and pollen) within a short distance compared to the ones traveling in forests. Indeed, Tibesigwa et al. (2019) noticed a positive association between wild bee pollination and agricultural productivity. Moreover, Tibesigwa et al. (2019) noted a decline in economic gains by 29% among farming households in Arusha due destruction of habitats for wild pollinators.

6 Conclusion and Recommendations

Given the national enabling investment environment, well-articulated policies and legal framework, and physical and soft infrastructure, it is anticipated that the commercialization of managed honey bees' pollination services is possible in Tanzania. The URT (2021a) noted that Tanzania is a hub of 9.2 million bee colonies capable of producing 138 metric tons of honey annually. Tanzania has 48 million ha of forests and woodlands suitable for beekeeping activities (Guyo and Legesse 2015; Kagya 2014). The tropical forest occupies 55% of Tanzania's area of 480,000 km^2 (Michael et al. 2019). The dry and Zambezian woodlands alone account for two-fifths of the national land area. The dry Miombo woodland extends from southern Lake Victoria to the Mozambique border. Meanwhile, the Zambezian Miombo forests lie along the Tanzanian western part of the Burundi border to the Malawi border while the Itigi-Sumbu thicket is found in central Tanzania.

The beekeeping sub-sector generates US$ 2 million per annum and employs almost 2 million dwellers along the bee products value chain (Tutuba and Vanhaverbeke 2018). In fact, beekeeping is an alternative source of income for people living in rural settings that is affordable to poor households as it requires relatively few resources (start-up capital) to commence production (Mwakatobe et al. 2021; Tutuba and Vanhaverbeke 2018). Beekeeping contributes about 33% of the total income of households in the Miombo woodlands of Tanzania (Mwakatobe et al. 2021). Yet, the beekeeping sub-sector has not been utilized to its full potential. Currently, Tanzania's natural honey and bee wax extraction are estimated to stand to the tune of only 34,000 tons and 1830 tons, which account for only 34% and 27% of production potential, respectively (URT 2021a). The opportunity of renting out the pollination services has not yet reaped its full potential. Essentially, the prevailing distribution of tropical forestry in the country, beekeeping activities, and thereby, managed pollinated services is possible for all maize-growing agroecological regions.

The Tanzanian government is committed to promoting beekeeping and managed honey bees' pollination services. Nine gazetted bee reserves and forests are declared as national, local government authority, and village and private bee reserves. The URT (2021a) identified the presence of 1280 beekeeper groups and 59 registered cooperatives engaged in beekeeping country-wide. The government's commitment to the beekeeping sector further is noticed by operating four government beekeeping demonstration centers strategically located in Manyoni (Singida), Kondoa (Dodoma), Handeni (Tanga), and Kibondo districts in the Kigoma region (URT 2021a). Establishing 181 apiaries in seven beekeeping zones by the government of Tanzania through Tanzania Forest Services Agency (TFS) is another milestone in commercializing the beekeeping sector that enhances the smooth provision of managed honey bees' pollination services.

Moreover, experts in the country are another added advantage to honey beekeeping ventures and managed pollination services. The experts in the beekeeping sector are baked in four training institutes: the Sokoine University of Agriculture, the University of Dar es Salaam, the Forest Training Institute and Beekeeping Training Institutes. The graduate from the four training institutes amounts to 250 graduates per annum (URT 2021a). These graduates can be employed in the beekeeping sector upon need. Moreover, the relatively low-wage bills in Tanzania, amounting to US$ 1200 per annum for an employee, are an added advantage (Haji 2015; URT 2021a).

The mainstreamed governance system embracing biodiversity conservation and beekeeping at the national and local government level entices investors in the sector. Well-articulated policies and regulatory frameworks from the central government to the village authorities are other advantages for investment in managed honey bees' pollination services.

There is a need to invest in managed honey bees' pollination due to the projected increased food demand resulting from the increase in national population and declining wild pollinators. The current struggle for farmers to increase maize productivity is by employing good agronomic practices, including using improved seeds, application of organic and inorganic fertilizers, and timely planting, weeding, and pesticide application. Renting of managed honey pollination services from honey bees is seen within a tiny proportion (0.2%) of farming households (168 out of 7,837,405) in 2019/2020 in Tanzania (URT 2022).

Evidence in Tanzania regarding improvement of agricultural production is mainly in the horticultural, pulses, and oil seeds sub-sectors. Yet, the increased crop yield has been evaluated based on wild pollination services whose habitats are degraded as the population increase. Therefore, it is high time we test the use of managed honey bees' pollination services in maize farms to expand the knowledge in search of alternative means of increasing farmland productivity for maize, a staple food in the country.

References

Abass AB, Ndunguru G, Mamiro P, Alenkhe B, Mlingi N, Bekunda M (2014) Post-harvest food losses in a maize-based farming system of semi-arid savannah area of Tanzania. J Stored Prod Res 57:49–57. https://doi.org/10.1016/j.jspr.2013.12.004

Adam RI, Quinhentos ML, Muindi P, Osanya J (2020) Beyond maize production: gender relations along the maize value chain in Tanzania. Journal of Gender, Agriculture and Food Security 5(2):27–41. https://doi.org/10.1177/0030727019888661

Arndt C, Farmer W, Strzepek K, Thurlow J (2012) Climate change, agriculture and food security in Tanzania. Rev Dev Econ 16(3):378–393. https://doi.org/10.1111/j.1467-9361.2012.00669.x

Bagachwa MS, Naho A (1994) A review of recent developments in the second economy in Tanzania. In AERC special paper 16, African economic research consortium, Nairobi (issue May). African Economic Research Consortium

Baregu ML (1987) The paradox of the Arusha declaration. A Journal of African Politics, Development and International Affairs 14(1):1–12

Barreiro Hurle J (2012) Analysis of incentives and disincentives for maize in the united republic of Tanzania (issue Oct)

Beghin J (2015) Assessing food security in Tanzania in the next decade. Agricultural Policy Review, Winter 201(3):3

Benjaminsen TA, Alinon K, Buhaug H, Buseth JT (2012) Does climate change drive land-use conflicts in the Sahel? J Peace Res 49(1):97–111. https://doi.org/10.1177/0022343311427343

Beyene AD, Kassie M (2015) Speed of adoption of improved maize varieties in Tanzania: an application of duration analysis. Technol Forecast Soc Chang 96:298–307. https://doi.org/10.1016/j.techfore.2015.04.007

Bongole A (2021) Climate smart agricultural practices and food security: a case of Mbeya and Songwe regions in Tanzania

CBD (2018) Decision adopted by the conference of the parties to the convention on biological diversity (CBD). Conservation and sustainable use of pollinators. Conference of the parties to CBD/COP/DEC/14/6. Conference of the Parties to the Convention on Biological Diversity 23(November):1–16

Charles H, Godfray H, Garnett T (2014) Food security and sustainable intensification. Philosophical Transactions of the Royal Society b: Biological Sciences 369(1639):6–11. https://doi.org/10.1098/rstb.2012.0273

Cochrane N, D'Souza A (2015) Measuring access to food in Tanzania: a food basket approach. Economic Information Bulletin (issue 135)

Elisante F, Ndakidemi P, Arnold SEJ, Belmain SR, Gurr GM, Darbyshire I, Xie G, Stevenson PC (2020) Insect pollination is important in a smallholder bean farming system. PeerJ 8:1–22. https://doi.org/10.7717/peerj.10102

Epule TE, Chehbouni A, Dhiba D (2022) Recent patterns in maize yield and harvest area across Africa. Agronomy 12(2). https://doi.org/10.3390/agronomy12020374

Erenstein O, Jaleta M, Sonder K, Mottaleb K, Prasanna BM (2022) Global maize production, consumption and trade: trends and R&D implications. Food Security0123456789. https://doi.org/10.1007/s12571-022-01288-7

Falanta E, Bengesi K (2018) Drivers and consequences of recurrent conflicts between farmers and pastoralists in Kilosa and Mvomero districts Tanzania. Journal of Sustainable Development 11(4):13. https://doi.org/10.5539/jsd.v11n4p13

Frederick B, Sabula L, Mruma S, Mzee F, Mtoka E, Masigo J, Ndunguru A, Swai E (2020) Maize production manual for smallholder farmers in Tanzania

Gallai N, Salles JM, Settele J, Vaissière BE (2009) Economic valuation of the vulnerability of world agriculture confronted with pollinator decline. Ecol Econ 68(3):810–821. https://doi.org/10.1016/j.ecolecon.2008.06.014

Garibaldi LA, Pérez-Méndez N, Garratt MPD, Gemmill-Herren B, Miguez FE, Dicks LV (2019) Policies for ecological intensification of crop production. Trends Ecol Evol 34(4):282–286. https://doi.org/10.1016/j.tree.2019.01.003

Giannini TC, Cordeiro GD, Freitas BM, Saraiva AM, Imperatriz-Fonseca VL (2015) The dependence of crops for pollinators and the economic value of pollination in Brazil. J Econ Entomol 108(3):849–857. https://doi.org/10.1093/jee/tov093

Giliba RA, Mpinga IH, Ndimuligo SA, Mpanda MM (2020) Changing climate patterns risk the spread of Varroa destructor infestation of African honey bees in Tanzania. Ecological Processes 9(1). https://doi.org/10.1186/s13717-020-00247-4

Godfrey N (2018) Anthropogenic and climatic factors affecting honey production: the case of selected villages in Manyoni District, Tanzania. Journal of Agricultural Biotechnology and Sustainable Development 10(3):45–57. https://doi.org/10.5897/jabsd2017.0292

Grote U, Fasse A, Nguyen TT, Erenstein O (2021) Food security and the dynamics of wheat and maize value chains in Africa and Asia. Frontiers in Sustainable Food Systems 4(February):1–17. https://doi.org/10.3389/fsufs.2020.617009

Guyo S, Legesse S (2015) Review on beekeeping activities, opportunities, challenges and journal of harmonized research in applied sciences. Journal of Harmonized Research in Applied Sciences, October, 15. www.johronline.com

Haji, M. (2015). Youth employment in Tanzania: taking stock of the evidence and knowledge gaps. International Development Research Centre 1–30

Hepelwa A, Selejio O, Mduma K (2013) The voucher system and the agricultural production in Tanzania: is the model adopted effective? Evidence from the panel data analysis. Environment for Development, August, 24

Kagya MA (2014) Beekeeping in Tanzania: country situation paper ministry of natural resources and tourism Tanzania beekeeping in Tanzania: country situation paper ministry of natural resources and tourism Tanzania content

Kassie M, Jaleta M, Mattei A (2014) Evaluating the impact of improved maize varieties on food security in rural Tanzania: evidence from a continuous treatment approach. Food Security 6(2):217–230. https://doi.org/10.1007/s12571-014-0332-x

Katinila N, Verkujil H, Mwangi W, Anandajayasekeram P, Moshi J (1998) Adoption of maize production technologies in southern Tanzania. Mexico, D.F.: International Maize and Wheat Improvement Center (CIMMYT), The United Republic of Tanzania, and The Southern Africa Centre for Cooperation in Agricultural Research (SACCAR) 20. https://doi.org/10.4172/2329-8863.1000151

Keskin I, Abdalla M (2019) The sociological anallysis of the failure of Ujamaa policy. ARHUSS 2(3):16

Khalifa SG, Alajmi MF, Zhao C, Masry SHD, Abdel-daim MM (2021) Crop production. Insects 688(12):1–23

Kihara J, Kizito F, Jumbo M, Kinyua M, Bekunda M (2020) Unlocking maize crop productivity through improved management practices in northern Tanzania. African Journal of Food, Agriculture, Nutrition and Development 20(7):17095–17112. https://doi.org/10.18697/ajfand.95.17965

Kikiwete J, Garang MG, Membe Y (2015) Evaluation of farming practices and environmental pollution in Manyara basin Tanzania. International Journal of Agricultural Sciences 5(6):864–877

Kim J, Mason NM, Mather D, Wu F (2019) The effects of the national agricultural input voucher scheme (NAIVS) on sustainable intensification of maize production in Tanzania. J Agric Econ 72(3):857–877. https://doi.org/10.1111/1477-9552.12431

Kleijn D, Winfree R, Bartomeus I, Carvalheiro LG, Henry M, Isaacs R, Klein AM, Kremen C, M'Gonigle LK, Rader R, Ricketts TH, Williams NM, Lee Adamson N, Ascher JS, Báldi A, Batáry P, Benjamin F, Biesmeijer JC, Blitzer EJ et al (2015). Delivery of crop pollination services is an insufficient argument for wild pollinator conservation. Nature Communications 6(May 2015). https://doi.org/10.1038/ncomms8414

Kumari E (2018) Maize production constraints in Tanzania

Lobulu J, Shimelis H, Laing M, Mushongi AA (2019) Maize production constraints, traits preference and current Striga control options in western Tanzania: farmers' consultation and implications for breeding. Acta Agric Scand Sect B Soil Plant Sci 69(8):734–746. https://doi.org/10.1080/09064710.2019.1652680

Luzi-Kihupi A, Killenga K, Bonsi C (2015) A review of maize, rice, tomato and banana research in Tanzania. Tanzania Journal of Agricultural Sciences 14(1):1–20

Lyimo NG (2005) Improving farmers' access to and management of disease resistant cultivars in the Southern Highlands of Tanzania. 8220(Za 0535), 96

Macauley H, Ramadjita T (2015) Cereal crops: rice, maize, millet, sorghum, wheat. University of Cape Coast Institutional Repository

Magehema A, Chang'a L, Mkoma S (2015) Implication of rainfall variability on maize production in Morogoro, Tanzania. International Journal of Environmental Sciences4(5):10

Magrini E, Vigani M (2016) Technology adoption and the multiple dimensions of food security: the case of maize in Tanzania. Food Security 8(4):707–726. https://doi.org/10.1007/s12571-016-0593-7

Mather D, Waized B, Ndyetabula D, Temu A, Minde I (2016) The profitability of inorganic fertilizer use in smallholder maize production in Tanzania: implications for alternative strategies to improve smallholder maize productivity. GISAIA/Tanzania Working Paper #4, June 2016, Tanzania working paper #4, 1–45

Michael M, Josh H, Richard K, Daudi M, Geofrey M, Hussein S (2019) Biodiversity and tropical forestry (FAA 118/119) assessment

Mkonda M (2022) The influence of soil organic carbon and climate variability on crop yields in Kongwa district, Tanzania. Environmental Management

Msongaleli BM, Rwehumbiza F, Tumbo SD, Kihupi N (2015) Impacts of climate variability and change on rainfed sorghum and maize: implications for food security policy in Tanzania. J Agric Sci 7(5):124–142. https://doi.org/10.5539/jas.v7n5p124

Mwakatobe A, Ntalwila J, Lesio N, Leweri C, Haule L, Kilimba N (2021) Challenges and opportunities of beekeeping in Mvomero district, Morogoro region, Tanzania. Journal of Entomology and Zoology Studies 9(2):1–5. https://www.entomoljournal.com/archives/2021/vol9issue2/PartA/8-3-375-403.pdf

Mwakisisile AJ, Mushi AR (2019) Mathematical model for Tanzania population growth. Tanzania Journal of Science 45(3):346–354. www.ajol.info/index.php/tjs/

Nkonya E, Xavery P, Akonaay H, Mwangi W, Anandajayasekeram P, Verkuijl H, Martella D, Moshi A (1998) Maize production adoption of maize

Nuss ET, Tanumihardjo SA (2010) Maize: a paramount staple crop in the context of global nutrition. Comprehensive Reviews in Food Science and Food Safety 9(4):417–436. https://doi.org/10.1111/j.1541-4337.2010.00117.x

Nyaligwa, Masuki L, Karwani G (2022). Variation in nutrient use efficiency under varying soil conditions in maize growing areas in northern zone, Tanzania—East Africa. International Journal of Current Research, 1

Pauw K, Thurlow J (2011) Agricultural growth, poverty, and nutrition in Tanzania. Food Policy 36(6):795–804. https://doi.org/10.1016/j.foodpol.2011.09.002

Rweyemamu DR (2019) Lincoln university digital thesis agropastoralist and farming communities in Morogoro region

Sawe T, Nielsen A, Eldegard K (2020) Crop pollination in small-scale agriculture in Tanzania: household dependence, awareness and conservation. Sustainability (Switzerland) 12(6):1–13. https://doi.org/10.3390/su12062228

Suleiman R, Rosentrater K (2015) Current maize production, postharvest losses and the risk of mycotoxins contamination in Tanzania. American Society of Agricultural and Biological Engineers Annual International Meeting 2015 4:3289–3414. https://doi.org/10.13031/aim.20152189434

Tarnavsky E, Chavez E, Boogaard H (2018) Agro-meteorological risks to maize production in Tanzania: sensitivity of an adapted water requirements satisfaction index (WRSI) model to rainfall. International Journal of Applied Earth Observation and Geoinformation 73(December 2017):77–87. https://doi.org/10.1016/j.jag.2018.04.008

Temu A, Ashimogo G (1999) Economic reform in Tanzania: impacts on maize production, marketing, prices and food security, p 8

Temu A, Manyama A, Mgeni C, Langyintuo A, Waized B (2011) Characterization of maize producing households in Manyoni and Chamwino districts in Tanzania. In: Country report—Tanzania. CIMMYT, Nairobi

The Montpellier Panel (2013) Sustainable intensification: a new paradigm for African agriculture. In: London: agriculture for impact. https://workspace.imperial.ac.uk/africanagriculturaldevelopment/Public/MontpellierPanelReport2013-SustainableIntensification-ANewParadigmforAfricanAgriculture.pdf

Tibesigwa B, Siikamäki J, Lokina R, Alvsilver J (2019) Naturally available wild pollination services have economic value for nature dependent smallholder crop farms in Tanzania. Sci Rep 9(1):1–10. https://doi.org/10.1038/s41598-019-39745-7

Tutuba N, Vanhaverbeke W (2018) Beekeeping in Tanzania: why is beekeeping not commercially viable in Mvomero? Afrika Focus 31(1):213–239. https://doi.org/10.21825/af.v31i1.9047

Tutuba NB, Msamula JS, Tundui HP (2019) Business model innovation for sustainable beekeeping in Tanzania: a content analysis approach. American Journal of Management 19(1). https://doi.org/10.33423/ajm.v19i1.1340

United Republic of Tanzania (URT) (1998) National beekeeping policy 1998, p 56

United Republic of Tanzania (URT) (2001) Agricultural sector development strategy I

United Republic of Tanzania (URT) (2004) The environmental management act, 2004

United Republic of Tanzania (URT) (2013) National agriculture policy 2013

United Republic of Tanzania (URT) (2015) Agricultural sector development strategy-II (issue Sept 2015)

United Republic of Tanzania (URT) (2016) The united republic of Tanzania 2014/15 annual agricultural sample survey. https://www.nbs.go.tz/nbs/takwimu/Agriculture/Annual_Agricultural_Sample_Survey_Report2014_15.pdf

United Republic of Tanzania (URT) (2018) National population projections

United Republic of Tanzania (URT) (2021a) 2020 Tanzania in figures (Issue June)

United Republic of Tanzania (URT) (2021b) National beekeeping policy implementation strategy (2021b–2031) (Issue May)

United Republic of Tanzania (URT) (2021c) National Environmental Policy, 2021c

URT (2015) Tanzania intended nationally determined contributions (INDC) (issue 2007). http://www4.unfccc.int/submissions/INDC/PublishedDocuments/Turkey/1/The_INDC_of_TURKEY_v.15.19.30.pdf

URT (2022) Guideline for management and use of honeybee colonies for pollination services in Tanzania (issue May)

WB (2018) Agricultural pollution. Field burning. World Bank Group, 1–7. https://documents.worldbank.org/curated/en/989351521207797690/pdf/124342-repl-WB-Knowledge-Burning.pdf%0A,https://openknowledge.worldbank.org/bitstream/handle/10986/29504/124342-BRI-p153343-PUBLIC-march-22-9-pm-WB-Knowledge-Burning.pdf?sequence=1&isAllowed=

Wilson RT, Lewis I, Guideline P, Amosun JO, Aluko OA, Adeniyan ON, Adamu U, Mrema J, Msaky J, Price AJ, Kelton JA, Katinila N, Verkuijl H, Mwangi W, Anandajayasekeram P, Moshi AJ, Mohammed Amin AT, Sabas P, Amos Alexander D et al (2015) Evaluation of EM technology on maize (Zea mays L.) growth, development and yield in Morogoro Tanzania research report prepared by. Advances in Crop Science and Technology 02(October):1–8. https://doi.org/10.4172/2329-8863.1000151

Wu F, Guclu H (2013) Global maize trade and food security: implications from a social network model. Risk Anal 33(12):2168–2178. https://doi.org/10.1111/risa.12064
Xiong W, Tarnavsky E (2020) Better agronomic management increases climate resilience of maize to drought in Tanzania. Atmosphere 11(9). https://doi.org/10.3390/atmos11090982
Zibaee A (2013) Rice: importance and future. Rice Research: Open Access 1(2):2013. https://doi.org/10.4172/jrr.1000e102

Decomposing Energy-Related CO_2 Emissions in Tunisia Using the LMDI Approach

Sana Essaber

1 Introduction

Global warming caused by greenhouse gas (GHG) emissions, mainly carbon dioxide (CO_2), has reached alarming levels (IEA 2017), justifying thus the call of the Paris Agreement to unify all countries' efforts to mitigate climate change and keep temperature rise well below 2 °C and preferably 1.5 °C compared to the pre-industrial era (UNFCCC[1] 2015). Since 2007, the Intergovernmental Panel on Climate Change (IPCC 2007) report has warned that energy use is the main source of pollutant emissions at the local, national, and global levels. According to this report, the projected energy-induced CO_2 emissions in 2030 will be 40–110% higher than in 2000. Then and since the emissions mainly result from the consumption of fossil fuels, reducing energy consumption seems to be the obvious remedy for this emissions problem. However, due to the possible negative impacts on economic growth, cutting back energy use is likely to be the "less traveled road" (Soytas and Saria 2009). Emissions reduction is essential to help meet government targets to limit climate change. However, examining changes in CO_2 emissions or energy use over time does not offer any insight into the reasons for these changes. Index decomposition analysis (IDA) methods are a largely accepted analytical instrument in policymaking to assess energy conservation programs (Reddy and Ray 2010). Decomposition analysis (Ang and Zhang 2000) can split changes in energy-induced CO_2 emissions over time into several different factors. This enables a better understanding of the reasons for the observed changes using a detailed accounting of changes in emissions. Factors behind

[1] United Nations Framework Convention on Climate Change.

S. Essaber (✉)
Higher Institute of Accounting and Business Administration (ISCAE), University of Manouba, Manouba, Tunisia
e-mail: sana.essaber@iscae.uma.tn

LEGI—Polytechnic School of Tunisia, University of Carthage, Carthage, Tunisia

M. El Amine Abdelli and M. Shahbaz (eds.), *Digital Economy, Energy and Sustainability*, Green Energy and Technology, https://doi.org/10.1007/978-3-031-22382-2_8

Table 1 Tunisian commitment toward UNFCCC

Year	Purpose
1992	Signature of the United Nations Framework Convention on Climate Change
1993	Ratification of the United Nations Framework Convention on Climate Change
2001	Submission of the first national communication
2003	Ratification of the Kyoto Protocol
2010	Submission of the GHG mitigation actions in response to the Copenhagen Agreement
2012	GHG inventory in Tunisia
2014	Submission of the second national communication
2014	Submission of the first two-year report
2015	Submission of the INDC
2016	Submission of the second two-year report
2017	Ratification of the Paris Agreement
2021	Submission of the update INDC

the change in production, demographic development, structure, energy intensity, and carbon emissions can be brought together in a framework that allows relating changes in each factor to changes in total emissions. Ang and Zhang (2000) provide a review of worldwide index decomposition analysis (IDA) studies up to the year 2000. Some studies cover energy demand rather than carbon emissions, and some only investigate the first three effects, like the study of Essaber (2015) in Tunisia.

Although Tunisia is the least contributor to global carbon emissions, with 3 tons of t CO_2e per capita in 2012 (according to the latest available national survey of GHG emissions of 2012), it has always been committed to a low-carbon development policy since the establishment of the UNFCCC in 1992.[2] Tunisia is committed to the international process against climate change (Kyoto Protocol, INDC,[3] Paris Agreement, etc.) and has consistently honored its promise to monitor and report on its progress in climate policy, as shown in Table 1.

The current study aims to decompose energy-related carbon emissions of both economic and residential sectors by adopting detailed sector-specific raw data. We aim to identify the factors that contribute the most to the change in Tunisian energy-induced carbon emissions from 1990–2012. The rest of this paper is structured as follows. Section 2 reviews the literature. Section 3 presents the methodology and data. Section 4 reports and discusses the results, while Sect. 6 concludes the paper.

---

[2] https://unfccc.int/sites/default/files/resource/docs/2012/cop18/eng/08a03.pdf.

[3] Intended nationally determined contribution.

2 Literature Review

Among the available methods, the log-mean Divisia index (LMDI) method developed by Ang (2005) is used to figure out the main factors contributing to the change in CO_2 emissions in Tunisia. Thus, having a better understanding of energy-induced emissions by all sectors, how they change, and the sources of these changes would have important policy implications in reducing CO_2 emissions and reaching the 2 °C and even the 1.5 °C global target. Many studies have used decomposition analysis on pollutant emissions for selected groups of countries. Bhattacharyya and Ussanarassamee (2004) used LMDI to decompose changes in aggregate energy intensities and CO_2 intensities in its multiplicative form for Thai industries between 1981 and 2000. The subsectoral investigation shows that the food and nonmetal industries were important factors in determining these intensities. Bacon and Bhattacharya (2007) briefly reviewed several of these studies and provided a decomposition of CO_2 emissions from the energy sector for 70 countries with the highest emissions levels. Their decomposition analysis of a period stretching from 1994 to 2004 found a relationship between the evolution of emissions during the decade and changes in several determinants. For the studied countries, GDP per capita was the dominant variable explaining the increase in total emissions, while the population was only about half as important.

Nevertheless, the decline in energy intensity was so significant that it balanced out approximately 40% of the joint impact of these two variables. In 2005, Wang et al. used LMDI to analyze China's change in aggregated CO_2 from 1957 to 2000. The study reveals that China has significantly lowered its CO_2 emissions, mainly thanks to improved energy intensity. In the same way, fuel shifts and the use of renewable energies were also behind the decrease in CO_2 emissions. In 2006, Ediger and Huvaz used LMDI to assess sectoral energy consumption in the Turkish economy from 1980 to 2000, a period known for significant changes in the economic and demographic structure of the country. Hatzigeorgiou et al. (2008) applied the arithmetic mean Divisia index (AMDI) and the logarithmic mean Divisia index (LMDI) techniques to decompose energy-induced CO_2 emissions in Greece from 1990 to 2002 into four factors: income effect, energy intensity effect, fuel share effect, and population effect. Period-wise and time series analyses show that the biggest contributor to the rise in CO_2 emissions in Greece is the income effect. Timilsina and Shrestha (2009) employed LMDI to determine the factors behind the growth of the transport sector's CO_2 emissions in 20 Latin American and Caribbean (LAC) countries during the 1980–2005 period. The key insight of the study is that economic growth and changes in transport energy intensity are the main contributors to the rise in carbon dioxide emissions in selected economies. De Freitas and Kaneko (2011) developed a decomposition model using a log-mean Divisia index (LMDI) framework to identify the determinants of emissions change in Brazil between 2004 and 2009. They found that carbon intensity and energy mix are the main factors behind a decrease in these emissions.

Similarly, O'Mahony et al. (2012) used the LMDI decomposition analysis method to identify and analyze the factors behind CO_2 emissions in final energy-consuming sectors in Ireland from 1990 to 2007. Their multi-sectoral analysis focuses on four economic sectors and the residential and transport sectors. Scale, structure, and intensity effects are also examined, showing significant sector performance heterogeneity.

3 Methodology and Data

3.1 *Data*

In this study, energy refers to the whole energy consumption for each final energy-use sector in the Tunisian economy without Biomass. The industries under study are agriculture, transport, tertiary, and residential sectors. All historical data (1990–2012) on these sectors' carbon emissions and energy consumption were collected from the 2012 national survey of GHG emissions and the National Energy Management Agency (NEMA). GDP, value-added, and population data in Tunisia were taken from the National Statistics Institute (NSI). We transformed the output and value data in constant local prices per 2015.

3.2 *Methodology*

The decomposition technique undertaken in this study is the index decomposition analysis (IDA) and is based on statistical data. It is used because of its simplicity and because the data are available. The IDA allows for several different variant techniques. A useful guide to these variants is given by Ang (2004). The log-mean Divisia index method (LMDI) was first introduced by Ang et al. (1998). Its basic mathematical formulae have been detailed in several studies (Ang and Zhang 2000; Ang and Liu 2001; Ang et al. 2003; Ang 2004, 2005). The LMDI method does not bear a residual term, and all zeros in the dataset may be replaced by a small positive constant (Lenzen 2006; Liu et al. 2007). Because of these two properties, we prefer this method. Additive decomposition is used in this study, where the difference in the factor investigated over a time period is decomposed into contributing factors. An alternative method is known as multiplicative decomposition, where the ratio rather than the difference is decomposed, and the additive approach was favored here for ease of interpretation.

In this paper, the log-mean Divisia index (LMDI) method developed by Ang (2005) is used to determine the main factors contributing to changes in CO_2 emissions in the Tunisian economy. In this paper, energy-induced carbon emissions can be calculated by the following formula where j refers to sectors:

$$C = \sum_j C_j \tag{1}$$

Since electricity consumption does not cause carbon emissions, we assume electricity's carbon emission coefficients are 0.

This study considers the industrial, agricultural, transport, and tertiary sectors separately from the residential sector. Since the transport sector is a major emitter of pollutants, NEMA considers it separately from the market services sector, as presented in national accounts.

Changes in CO_2 emissions in the four economic sectors in Tunisia (agriculture, industry, transport, and tertiary sectors) between two years (t) and ($t - 1$) between 1990 and 2012 are a sum of the changes due to changes in activity (ΔC_{Act}), changes in inter-sector structure (ΔC_{Str}), changes in energy intensity (ΔC_{Int}), and changes in emission factors (ΔC_U).

Changes in CO_2 emissions in the residential sector can be decomposed into the population effect (ΔC_{pop}), energy intensity effect (ΔC_{IE}), energy mix effect (ΔC_{EM}), and CO_2 emission factor effect (ΔC_U).

Moreover, when referring to changes, this study will use the log percentage change ($L\%$) instead of the ordinary percentage because the latter has non-additive and asymmetric properties (Tornqvist and Vartia 1985). The relative changes of X_1 and X_2 are as follows:

$$L\% = \ln\left(\frac{X_1}{X_2}\right) \times 100 = \left[\frac{(X_2 - X_1)}{L(X_2, X_1)}\right] \times 100$$

It indicates that the log difference is relative to the logarithmic meaning. $L\%$ is symmetric, in effect independent of which point is taken as a comparison point; then, additive relative changes can be added successively and normally. All these properties are desirable for measuring relative changes (Tornqvist and Vartia 1985).

For j sectors of the Tunisian economy, total carbon emissions can be given by:

$$C = \sum_j \left(\text{GDP} \times \frac{VA_j}{\text{GDP}} \times \frac{E_j}{VA_j} \times \frac{C_j}{E_j}\right) + \left(P \times \frac{E_{res}}{P} \times \frac{C_{res}}{E_{res}}\right)$$

$$C = \sum_j \text{GDP} \times S_j \times IE_j \times U_j + (P \times IE_{res} \times U_{res}) \tag{2}$$

where C represents energy-induced carbon emissions of the five final energy-use sectors: agriculture, industry, transport, tertiary, and residential; GDP represents the gross domestic product; VA_j represents the output value of each of the agriculture, industry, transport, and tertiary sectors; E_j represents energy consumption of the final energy-use sector j; C_j represents the carbon emissions caused by the economic sectors; S_j (economic structure) represents the proportion of GDP accounted for by the output value of the economic sector j; IE_j represents energy intensity of the economic sector j; U_j represents the energy-carbon effect of the economic sector

j. For the residential sector, P represents the total population; IE_{res} represents residential energy consumption per capita; U_{res} represents the energy-carbon effect of the residential sector. Accordingly, we use the following notations: 'Act' for the activity effect, 'S' for the structure effect, '*IE*' for the energy intensity effect, '*U*' for energy-carbon effect, and '*P*' for the population effect.

Changes in carbon emissions between any two years (t) and ($t - 1$) are decomposed as follows:

$$\begin{aligned} \Delta C = C_t - C_{t-1} = \Delta C_{\text{Act}} + \Delta C_{S_j} + \Delta C_{IE_i} + \Delta C_{U_j} \\ + \Delta C_P + \Delta C_{IE_{\text{res}}} + \Delta C_{U_{\text{res}}} \end{aligned} \tag{3}$$

According to the additive LMDI formula, each effect on the right side of Eq. 3 can be written as follows:

$$\Delta C_{\text{Act}} = \sum_j \frac{C_j^t\_C_j^{t-1}}{\ln C_j^t - \ln C_j^{t-1}} \ln\left(\frac{\text{GDP}^t}{\text{GDP}^{t-1}}\right) \tag{4}$$

$$\Delta C_{S_j} = \sum_j \frac{C_j^t\_C_j^{t-1}}{\ln C_j^t - \ln C_j^{t-1}} \ln\left(\frac{S_j^t}{S_j^{t-1}}\right) \tag{5}$$

$$\Delta C_{IE_j} = \sum_j \frac{C_j^t\_C_j^{t-1}}{\ln C_j^t - \ln C_j^{t-1}} \ln\left(\frac{IE_j^t}{IE_j^{t-1}}\right) \tag{6}$$

$$\Delta C_{U_j} = \sum_j \frac{C_j^t\_C_j^{t-1}}{\ln C_j^t - \ln C_j^{t-1}} \ln\left(\frac{U_j^t}{U_j^{t-1}}\right) \tag{7}$$

$$\Delta C_P = \sum \frac{C^t_{\text{res}-C_{\text{res}}^{t-1}}}{\ln C_{\text{res}}^t - \ln C_{\text{res}}^{t-1}} \ln\left(\frac{P^t}{P^{t-1}}\right) \tag{8}$$

$$\Delta C_{IE_{\text{res}}} = \sum \frac{C_{\text{res}-}^t C_{\text{res}}^{t-1}}{\ln C_{\text{res}}^t - \ln C_{\text{res}}^{t-1}} \ln\left(\frac{IE_{\text{res}}^t}{EI_{\text{res}}^{t-1}}\right) \tag{9}$$

$$\Delta C_{U_{\text{res}}} = \sum \frac{C_{\text{res}-}^t C_{\text{res}}^{t-1}}{\ln C_{\text{res}}^t - \ln C_{\text{res}}^{t-1}} \ln\left(\frac{U_{\text{res}}^t}{U_{\text{res}}^{t-1}}\right) \tag{10}$$

4 Results and Discussion

In this section, we examine the factors behind the increase of energy-induced CO_2 emissions in Tunisia between 1990 and 2012, both at global and sectoral levels, namely the five final energy-use sectors: industry, agriculture, tertiary, transport, and the residential sector.

4.1 Global Decomposition Analysis

Changes in energy-induced CO_2 emissions are decomposed into the following factors: the activity effect (ΔC_{Act}), the energy intensity effect (ΔC_{IE}), the structure effect (ΔC_S), the population effect (ΔC_{pop}), and the carbon emission effect (ΔC_U). The units are in kilo tons of CO_2 equivalent (ktCO_2e).

LMDI gives perfect decomposition, i.e., the results do not contain an unexplained residual term. The decomposition results (1990–2012) for the economy-wide and the five sectors in the study are depicted in Fig. 1.

The following results are reached: There is a continuous decline in both energy intensity (ΔCIE) and carbon emission effects (ΔC_U). This can be explained by a decline in the share of industrial production in the GDP as the share of services in the Tunisian GDP has grown since 2000. In addition, the increase in electricity and natural gas use and the decrease in the use of oil products generated a decrease in the aggregate energy intensity. It is also clear that the activity effect (ΔC_{Act}) and the energy intensity effect (ΔC_{IE}) have strongly contributed to CO_2 emissions changes. Moreover, the decomposition results reveal that the decline in energy intensity offsets the increase in economic activity during the studied period.

To better assess the factors behind changes in CO_2 emissions in Tunisia during the 1990–2012 period, we conduct a more disaggregated decomposition of the factors explaining variation in end-user energy-induced CO_2 emissions by sectors,

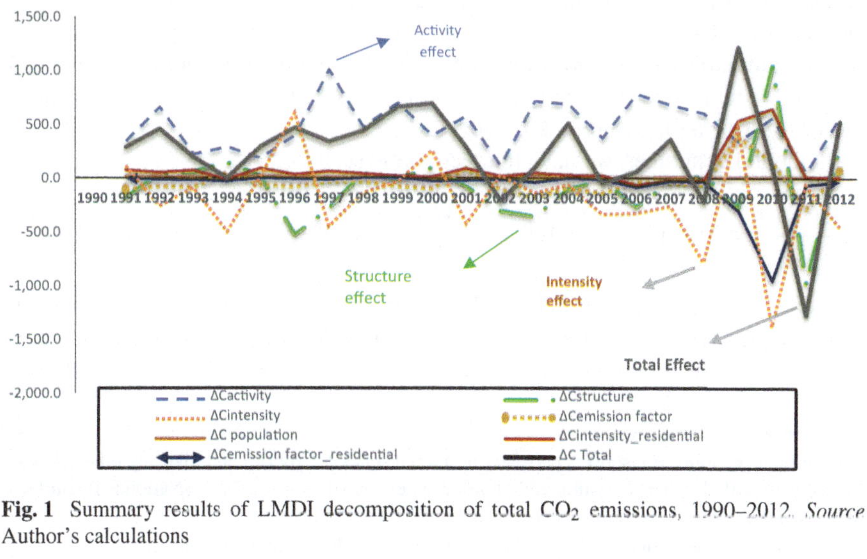

Fig. 1 Summary results of LMDI decomposition of total CO_2 emissions, 1990–2012. *Source* Author's calculations

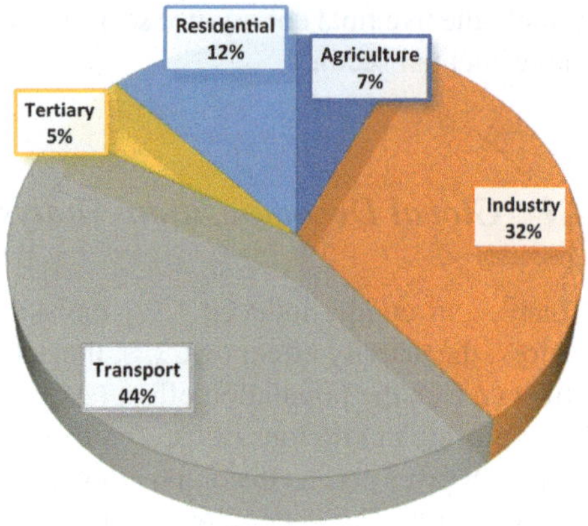

Fig. 2 Distribution of Tunisia's CO_2 emissions by final energy-use sector in 2012. *Source* NEMA

namely the economic sectors (industry, agriculture, transport,[4] and tertiary[5]) and the residential sector.

4.2 Additive LMDI Decomposition Results for CO_2 Emissions Growth in Agriculture, Industry, Transport, and Tertiary Sectors

Through an additive LMDI decomposition, we determined the factors behind the cumulated change in energy-induced CO_2 emissions in Tunisia's economic sector. Indeed, as illustrated in Fig. 2, the four national economic sectors, agriculture, industry, transport, and tertiary, accounted for 88% of total energy-induced CO_2 emissions in 2012.

The summarized decomposition results of the economic sector are shown in Fig. 3. The total accumulated effect on CO_2 emissions from the four economic sectors during the 1990–2012 period is 4634 $ktCO_2e$.

In the index number, we have four economic sectors:

$$\frac{\Delta C_{\text{Act}}}{\Delta C} \times 100\% + \frac{\Delta C_S}{\Delta C} \times 100\% + \frac{\Delta C_{IE}}{\Delta C} \times 100\% + \frac{\Delta C_U}{\Delta C} \times 100\% = 100\%$$

Table 1 reports the results of our calculations by economic sector and contributing effect. By sector, a quite revealing result is that the transport sector is the most contributor to the total effect with almost 70% of the total economy. It is followed

[4] In national accounts, transport is part of commercial service activities and, like any other activity, presents an added value. It is this added value given by official database (National Institute of Statistics) that is used in our decomposition calculations.

[5] The tertiary sector includes business, administration, hotels, communications, etc.

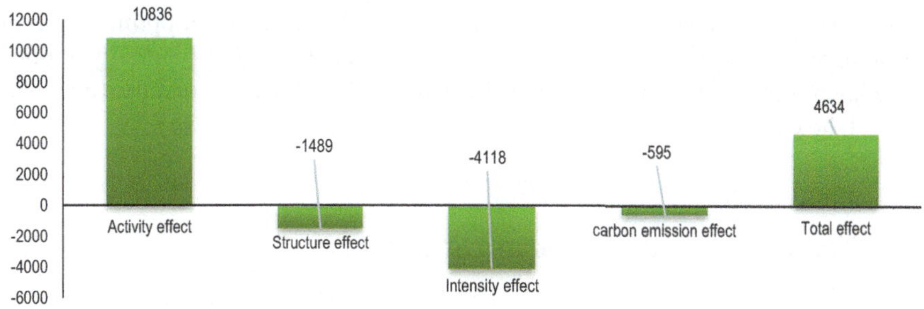

Fig. 3 Decomposition of CO_2 emission in economic sectors, 1990–2012, Cumulated effect results, in ktCO_2e. *Source* Author's calculations, NEMA data

by the industrial sector (17%) and then, with the same percentage (7%), by the other sectors, agriculture, and tertiary.

Economic activity is the dominant factor behind CO_2 emissions growth, with an increase of 10,836 ktCO_2e from 1990 to 2012, which accounts for 234% of total change. However, this activity impact was offset by a fall in intensity (−4118 ktCO_2e) and structure (−1489 ktCO_2e) effects during the studied period. The other influencing factor, the carbon emission effect, is negative and relatively small and accounts for only 13% of total change.

In what follows, we discuss each of the factors and their effect on CO_2 emissions.

4.3 The Influence of Economic Activity

Tunisia's GDP reached, in 2012, 36,844 million constant 2005 Tunisian dinars, growing at an average annual rate of 3.7% from 1990 to 2012. Economic growth was always a prioritized political goal in the country since the 1990s, while environmental protection was emphasized to optimize the economic structure and control the speed of final energy consumption and its effect on CO_2 emissions in the country. In fact, for the productive sectors, economic activity seems to be the dominating factor behind emissions growth (10,836 ktCO_2e). It was positive during the studied period (Fig. 2). This means that Tunisia's economic growth contributes directly to annual increases in CO_2 emissions.

Moreover, reducing GDP is not the solution even though rising incomes can make urban inhabitants' lifestyles more energy-intensive (Dhakal 2009). Increases in energy demand have become a fundamental issue for economic activity, and with such a high reliance on combustibles, CO_2 emissions have inexorably increased, particularly in transport and industry. It is worth mentioning that the transport activity and the resulting CO_2 emissions significantly parallel transport structure in Tunisia, with the predominance of the road transport system where gasoline and diesel are the main used fuels. Tax mechanisms such as public transport incentives, using renewable energy fuels, and launching cleaner vehicle sales will be useful in optimizing fuel consumption and shifting activities in the country. Reaching carbon neutrality implies a complete paradigm shift, requiring deep changes and a level of investment far beyond the current packages.

4.4 The Influence of Economic Structure

As depicted in Fig. 4, the GDP structure has changed, especially since 2000, with a share of GDP growth in favor of services first and industry second.

The positive coefficient of the structural change effect in the tertiary sector (151 $ktCO_2e$) reveals that the service sector's share in the global GDP increased at the expense of agriculture and industry (Table 2).

The negative coefficient of the structural change effect shows a decline in CO_2 emissions generated by a decline in the share of the industrial and agricultural sectors in the GDP. The same is true for the transport sector as the negative coefficient of the structural effect (−614 $ktCO_2e$) indicates that this sector contributes less to GDP, accounting for almost 44% of total national CO_2 emissions as reported by the 2012 national survey of GHG emissions (Fig. 2).

For the industrial sector, the second-largest value-added productive sector after services, it can be argued that the rapid economic growth during the 1990s was the main factor explaining energy demand and CO_2 emissions (Essaber 2015).

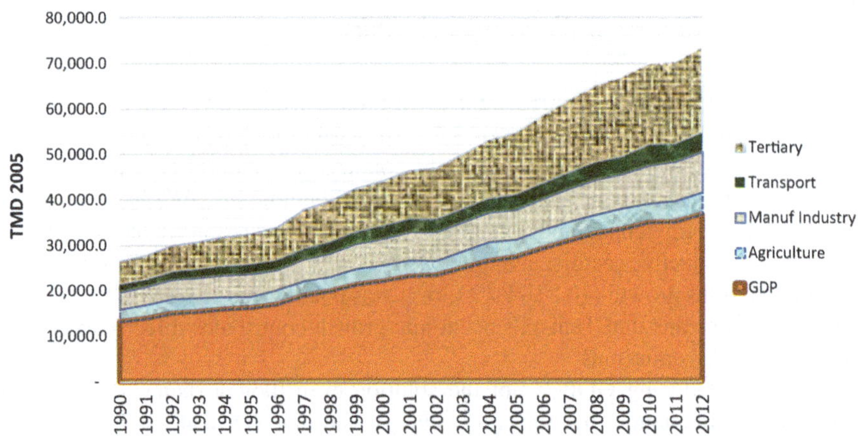

Fig. 4 GDP and value-added by productive sector, 1990–2012. *Source* National Institute of Statistics (INS)

Table 2 Summary of LMDI decomposition results of energy-induced CO_2 emissions in Tunisian economic subsectors, by sector and by contributing effect, 1990–2012, in $ktCO_2e$

Economic subsector	Activity level impact (GDP)	Structure level impact (VAj/GDP)	Intensity level impact (Ej/VAj)	Carbon emission effect (Cj/Ej)	Total	Share by sector (%)
Agriculture	1003	−468	−25	−186	324	**7**
Industry	4180	−557	−2315	−505	803	**17**
Transport	5062	−614	−1448	205	3204	**69**
Tertiary	591	151	−330	−108	303	**7**
Economy	10,836	−1489	−4118	−595	4634	**100**
Share in total effect	**234%**	**−32%**	**−89%**	**−13%**	**100%**	

Source Author's calculations, NEMA data

4.5 The Influence of Energy Intensity

Energy intensity measures the quantity of energy required per unit of GDP. It is a measure that reflects how energy use is efficient in an economy. A lower energy intensity reveals that more value-added can be generated with less energy inputs. During the 1980s, both primary and final energy intensities recorded a rise in Tunisia because of an accelerated growth rate of industrial sectors with highly intensive energy use, such as the IMCCV (construction materials, ceramic, and glass industries) and mines. Since the 1990s, these two intensities have started to decrease. As seen in Table 2, these intensities negatively contributed to overall CO_2 emissions, with a decrease of 4118 $ktCO_2e$, accounting for 89% of total CO_2 emissions in the accumulated effect. This can be attributed to a decrease in agriculture and tertiary intensities, yet it seems that industrial and transport energy intensities inhibited carbon emissions in the whole studied period, as shown in Table 2.

These results point to the positive effects of the country's strategies to improve energy efficiency. However, current policies are ineffective enough to significantly reduce energy demand and CO_2 emissions. Tunisia's installed renewable energy structure is modest when assessed against its large-scale solar photovoltaic and wind resources. Only, proactive scenarios with more ambitious measures of energy efficiency along with technology development could positively affect current trends and lead to significant effects on saving energy and reducing pollutant emissions.

4.6 The Influence of the Energy-Carbon Index

The energy-carbon index is a measure used to depict how low or high is carbon level in the energy mix. The LMDI decomposition results reveal that the energy-carbon effect, defined as carbon emissions by the unit of energy consumed in sector j, is negative and relatively small (−595), accounting for only 13% of the total effect during 1990–2012 in the Tunisian production sectors. It played a relatively minor role in decreasing CO_2 emissions revealing that fuels are still an important component compared to low-carbon sources in the energy mix. Moreover, the cumulative effect is positive for the transportation sector, indicating that it is a high CO_2-emitting sector because of its high fuel use.

5 Additive LMDI Decomposition Results for CO_2 Emissions Growth in the Residential Sector

The decomposition analysis of residential carbon dioxide emissions shows that energy intensity largely contributed to increasing CO_2 levels, followed by the population effect (Fig. 5).

The cumulated changes reported in Fig. 5 reveal that the quantity of CO_2 emissions generated by the residential sector increased by 654 $ktCO_2e$.

Our calculations show that total changes in CO_2 emissions in the Tunisian residential sector result, respectively, from positive changes in the population and intensity effects and negative changes in the carbon emission effect (Table 3).

In the index number, we have four residential sectors:

$$\frac{\Delta C_P}{\Delta C} \times 100\% + \frac{\Delta C_{IE_{\mathrm{res}}}}{\Delta C} \times 100\% + \frac{\Delta C_{U_{\mathrm{res}}}}{\Delta C} \times 100\% = 100\%$$

Indeed, our results reveal that the contribution of the population to CO_2 emissions is high. The cumulative effect increased to 431 $ktCO_2e$, accounting for 66% of the total change during the studied period. In fact, the growing population and rapid urbanization (almost 65% of the total population in Tunisia is urban) have significantly contributed to increasing CO_2 emissions in the country. However, the intensity-level effect, calculated as the amount of energy per capita, is the main contributor

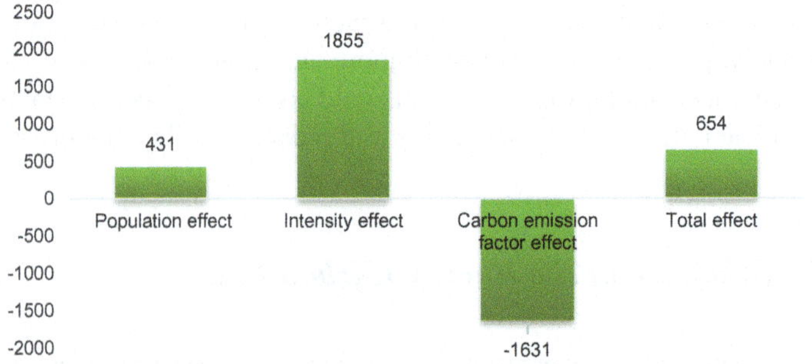

Fig. 5 Decomposition of CO_2 emissions in the residential sector ($ktCO_2e$), 1990–2012. *Source* Author's calculation, NEMA data

Table 3 Summary of LMDI decomposition results of energy-induced CO_2 emissions in the Tunisian residential sector, 1990–2012

	Population-level effect (P)	Intensity-level effect (E_{res}/P)	Carbon emission-level effect (C_{res}/E_{res})	Total
1990–2012 ($ktCO_2e$)	431	1855	−1631	654
Share in total effect (%)	66	283	−249	100

Source Author's calculations, NEMA data

to the increase in CO_2 emissions: The accumulated effect of energy intensity saw an increase of 1855 $ktCO_2e$ which accounts for 283% of the total change. Its positive coefficient mainly results from rising incomes and the continuous concentration of energy-consuming behavior in urban areas. Some measures are then necessary to reduce total emissions of carbon dioxide. However, decreasing population size cannot be considered since Tunisia already has a very low population growth rate. However, inducing and adopting new technology will improve energy efficiency and reduce energy use per capita. This can justify why environmental policies highly focus on reducing carbon dioxide per unit of used energy (Ang and Liu 2001). In Tunisia, the results reveal that the energy-carbon effect is important and negative in the residential sector, with an average annual contribution of 249%. This means that thanks to the policies taken, 1631 $ktCO_2e$ is no longer emitted. The introduction of non-fossil fuels in heat and electricity generation slowed CO_2 emissions growth (Ma et al. 2018).

6 Conclusions

It is useful for Tunisia's energy and environmental policy-makers to investigate the factors explaining CO_2 emissions levels and their progress. Identifying sector-specific factors affecting variation in energy-induced carbon dioxide emissions can be relevant in identifying suitable proactive actions aiming at responding effectively to greenhouse gas emission growth in the country. We used the log-mean Divisia index (LMDI) method developed by Ang (2005) to identify the factors behind changes in CO_2 emissions resulting from changes in activity, demography, activity structure, energy intensity, and emission factors at the global and sectoral levels during the 1990–2012 period. The results indicate that the main factors explaining the change in CO_2 emissions are changes in total activity and energy intensity. Pollutant emissions are the product of a very complex dynamic system determined by factors such as demographic growth, economic development, and technological change.

In general, the energy agenda in Tunisia during this period mainly focused on the energy efficiency (EE) strategy, relegating national strategies to promote renewable energy (RE). More efforts should be made through both EE and RE to reduce CO_2 emissions while ensuring sustainable development in the country. These efforts undoubtedly need governmental support. In this regard, and since the aim is to encourage a more de-carbonized energy future, a new wave of plans and measures are in place in the country. In addition to its development priorities, Tunisia can envision an important greenhouse gas mitigation project requiring human commitment and technical and financial resources (Initial communication of Tunisia 2001). In fact, according to the GHG Abatement Plan, mitigation options in the energy sector are divided into three categories: demand side management options (residential and tertiary, industry, and transport sectors), supply side management options (centralized electricity production and promotion of alternative energy supply system), and cross-cutting options. Only then could Tunisia meet the Paris Agreement's targets on climate change.

Acknowledgements The author thanks Professor B. Sudhakara Reddy, India, for critically reviewing the manuscript and providing valuable suggestions.

References

Ang BW (2004) Decomposition analysis for policymaking in energy: which is the preferred method? Energy Policy 32

Ang BW (2005) The LMDI approach to decomposition analysis: a practical guide. Energy Policy 33:867–871

Ang BW, Liu FL (2001) A new energy decomposition method: perfect in decomposition and consistent in aggregation. Energy 26:537–548

Ang BW, Zhang FQ (2000) A survey of index decomposition analysis in energy and environmental studies. Energy 25:1149–1176

Ang BW, Liu FL, Chew EP (2003) Perfect decomposition techniques in energy and environmental analysis. Energy Policy 31(14):1561–1566

Ang BW, Zhang F, Choi KH (1998) Factorizing Changes in Energy and Environmental Indicators through Decomposition. Energy 23(6):489–495

Bacon R, Bhattacharya S (2007) Growth and CO_2 emissions: how do different countries fare? Environment Department Paper, no. 113, Word Bank

Bhattacharyya SC, Ussanarassamee A (2004) Decomposition of energy and CO_2 intensities of Thai industry between 1981 and 2000. Energy Economics 26:765–781

De Freitas LC, Kaneko S (2011) Decomposition of CO_2 emissions change from energy consumption in Brazil: challenges and policy implications. Energy Policy 3(39):1495–1504

Dhakal S (2009) Urban energy use and carbon emissions from cities in China and policy implications. Energ Policy 37:4208–4219

Ediger VS, Huvaz O (2006) Examining the sectoral energy use in Turkish economy (1980–2000) with the help of decomposition analysis. Energy Convers Manage 47:732–745

Essaber JS (2015) Energy consumption in Tunisia over 1990–2008: a decomposition analysis using logarithmic mean Divisia index technique. In: Sudhakara Reddy B, Ulgiati S (eds) Energy security and development—the global context and Indian perspectives. Springer, pp 147–161

Hatzigeorgiou E, Polatidis H, Haralambopoulos D (2008) CO_2 emissions in Greece for 1990–2002: a decomposition analysis and comparison of results using the arithmetic mean Divisia index and logarithmic mean Divisia index techniques. Energy 33(3):492–499
International Energy Agency—IEA (2017) World energy outlook, OECD/IEA
Intergovernmental Panel on Climate Change (2007) Climate Change 2007: synthesis report. Switzerland, Geneva
Lenzen M (2006) Decomposition analysis and the mean-rate-of-change index. Appl Energy 83:185–198
Liu L, Fan Y, Wu G, Wei Y (2007) Using LMDI method to analyze the change of China's industrial CO_2 emissions from final fuel use: an empirical analysis. Energy Policy 35:5892–5900
Ma L, Chong C, Zhang X, Liu P, Li W, Li Z, Ni W (2018) LMDI decomposition of energy-related CO_2 emissions based on energy and CO_2 allocation Sankey diagrams: the method and an application to China. Sustainability 10:344. https://doi.org/10.3390/su10020344
Ministry of Environment and Land Planning (2001) Initial communication of Tunisia under the United Nations Framework Convention on Climate Change. Republic of Tunisia, 2001 Oct
National Institute of Statistic of Tunisia. See: http://www.ins.nat.tn/indexfr.php/
O'Mahony T, Zhou P, Sweeney J (2012) The driving forces of change in energy-related CO_2 emissions in Ireland: a multi-sectoral decomposition from 1990 to 2007. Energy Policy 44:256–267
Reddy BS, Ray BK (2010) Decomposition of energy consumption and energy intensity in Indian manufacturing industries. Energy Sustain Dev 14:35–47
Soytas U, Saria R (2009) Energy consumption, economic growth, and carbon emissions: challenges faced by an EU candidate member. Ecol Econ 68:1667–1675
Timilsina GR, Shrestha A (2009) Factors affecting transport sector CO_2 emissions growth in Latin American and Caribbean countries: an LMDI decomposition analysis. Int J Energy Res 33(4):396–414
Tornqvist L, Vartia P (1985) How should relative changes be measured? Am Stat 39(1):43–46
UNFCCC (2015) Adoption of the Paris agreement, conference of the parties twenty-first session Paris, 30 November to 11 December 2015
Wang C, Chen JN, Zou J (2005) Decomposition of energy-related CO_2 emission in China: 1957–2000. Energy 30:73–83

Soil Properties and Carbon Sequestration Using Biochar and Compost on an Alfisol in Southwest Nigeria

Bukola M. Aina, Abiola O. Saliu, and Moses B. Adewole

1 Introduction

Soil is known as a sink for carbon, but factors such as anthropogenic activities and land use changes could reduce soil's potential capacity to store carbon (Kushwah et al. 2019). Carbon is essential to the soil for enhanced water-retention capacity, structure and fertility (Lal 2004). Carbon stock in the soil also minimizes the buildup of carbon dioxide and methane (Elum and Momodu 2017; Lal 2019) in the atmosphere.

Carbon sequestration is capturing and storing carbon from the atmosphere in the ocean, soil, forest vegetation and geological formations over time (Olson 2013; Nunes et al. 2020). Carbon captured, stored and transferred from plant biomass to soil in the form of soil organic carbon (SOC) is in the labile pool (Leno 2020). Soil organic carbon storage has been widely considered as a measure for mitigating global climate change through C sequestration in soils (Lal 2004). Increased soil carbon sequestration can enhance soil quality due to carbon's importance in soil's physical, chemical and biological properties. A recent technological innovation to reduce atmospheric CO_2 concentration by sequestering it is biochar application to soil.

Biochar is an organic material produced from biodegradable biomass feedstocks such as wood, plant residues, animal excreta, coconut husk, green wastes and forestry residues and many more using pyrolysis technology (Yang et al. 2017; Akmal et al. 2019). Biochar is a material with low porous and surface area that can sequester and store atmospheric carbon when applied to soil (Liu et al. 2016; Olmo et al. 2016). The chars differ in physical and chemical compositions based on the biomass content, and thermal conversion process used (Jindo et al. 2014; Karbout et al. 2019). Biochar

B. M. Aina · M. B. Adewole (✉)
Institute of Ecology and Environmental Studies, Obafemi Awolowo University, Ile-Ife, Nigeria
e-mail: badewole@oauife.edu.ng

A. O. Saliu
Department of Plant and Environmental Biology, Kwara State University, Malete, Ilorin, Nigeria

M. El Amine Abdelli and M. Shahbaz (eds.), *Digital Economy, Energy and Sustainability*, Green Energy and Technology, https://doi.org/10.1007/978-3-031-22382-2_9

helps reduce the volume and weight of feedstocks and offers a soil amendment that enhances soil's physical and chemical properties (Singh et al. 2019).

Soil carbon sequestration could increase with cultural practices such as biochar, compost and manure addition to soil, maintain the level of greenhouse gas in the atmosphere and enhance plant growth (Chen et al. 2018). The loss and gain of SOC depend on soil type, temperature, erosion, vegetation type and land management techniques. Liang et al. (2006), Hardy et al. (2016) and Carrari et al. (2018) observed that the long-term impact of biochar on soil nutrient accessibility, particularly, is the enhancement of cation exchange capacity (CEC). Field experiments with biochar application by Macdonald et al. (2014) and El-Naggar et al. (2019) have equally established increased nutrient retention and enhanced crop yield. Also, Lal (2004) and Agegnehu et al. (2016) reported on the reduction in the enrichment of atmospheric greenhouse gases such as CO_2 and CH_4 when biochar is applied to the soil.

In Nigeria, due to easy accessibility to improved varieties and the growing increase of maize, it has become an important staple and cash crop for smallholder farmers (Alabi et al. 2018; Ogunniyan et al. 2019). The economic significance of maize includes biomass fuel from maize ethanol (Oladejo and Adetunji 2012), livestock forage and feeds (Dunham and Elaswad 2018) and varieties of industrial uses (Monday 2018). Despite all these uses of maize crops, factors such as unstable climatic conditions and soil infertility have been recognized as restricting maize crop output. Soils in Nigeria are generally kaolinitic with low SOC and CEC. High cropping intensity on the same piece of land is peculiar to resource-poor Nigerian farmers because the inability to procure and apply needed fertilizers worsens crop yield.

Organic substances, including maize cobs and maize stovers from preceding cultivation as plant materials and composts of cow dung and poultry droppings as animal materials, could be slowly pyrolysed to form biochar for soil quality enhancement of a field experiment. Presently, animal manure production is on the increase in Nigeria, and this litters the environment as the attention to turn them to biological nutrient sources has not been given adequate attention. This paper, therefore, assessed the effects of different types of biochar and compost on soil properties, growth response, and yield of maize crops over a 5-year period. It also assessed the rate of soil C sequestration in an Alfisol of the forest zone of Southwestern Nigeria.

2 Materials and Methods

2.1 Experimental Layout, Compost and Biochar Production

The field experiments were conducted at the Teaching and Research Farm (T&RF), Obafemi Awolowo University (OAU), Ile-Ife, Nigeria (Fig. 1) in the late maize cropping season (August–November) of 2014–2018. The soil of the study area was an Alfisols classified locally as Iwo series, and the experimental plots covered latitude

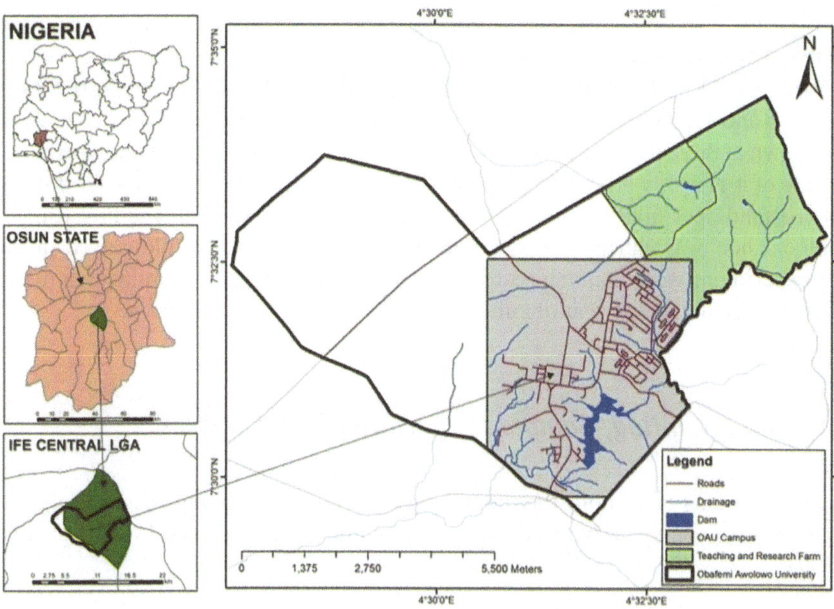

Fig. 1 Map of Nigeria showing the study area

07° 30′ and 07° 35′ N and longitude 04° 30′ and 04° 32′ E, 244 m above sea level. The mean annual rainfall at the experimental site was 1520.29 mm with a bimodal rainfall pattern; the rainy season started in March and ended in November, with a short dry spell in April and August.

Fresh cow dung and poultry droppings were collected from the Beef and Poultry Unit of T&RF, OAU, Ile-Ife. The two animal wastes were separately heaped under a shed and allowed to undergo the composting process aerobically for three months. During composting, the heaps of cow dung and poultry droppings were turned once a week to enhance the decomposition rate. Some portions of the cow dung and poultry droppings compost were further separately pyrolysed using a local charcoal-fired reactor designed for small-scale biochar production (Odesola and Owoseni 2010). Also, the plant wastes, maize cobs and maize stovers obtained from the T&RF, OAU, Ile-Ife, Nigeria were separately pyrolysed using the same method.

Land preparation was achieved by ploughing the field twice using a tractor in 2014 and subsequently with a handheld hoe each year. The experiment was carried out on a site measuring 29.0 m × 11.5 m, within which an alley of 1.0 m between blocks and 1.0 m within plots was established. There were seven experimental variables namely, maize cobs biochar (100% MCB), maize stovers biochar (100% MSB), cow dung biochar (100% CDB), poultry droppings biochar (100% PTB), equal combination of cow dung compost and biochar (50% CDC + 50% CDB), equal combination of poultry droppings compost and biochar (50% PTC + 50% PTB), and zero treatment application served as control. Each experimental variable was replicated three times to give a total of 21 plots, each measuring 3.0 m × 2.5 m and laid out in a randomized complete block design.

Variety ART 98/SW1 of maize seed purchased from the Institute of Agricultural Research and Training (IAR &T), Ibadan, Nigeria, was the test crop. The maize seeds were sown at three per hole using 75 cm × 50 cm sowing distance, and treatments, at the rate of 4 t ha^{-1}, were applied at sowing. The maize seedlings were later thinned to two stands per hole two weeks after sowing (WAS). Manual weeding using a handheld hoe was carried out at 2 and 5 WAS. The treatment plots used in the first experiment were maintained for subsequent ones, but with similar treatments addition only in the 3rd year of cultivation.

2.2 Soil Sampling, Sample Preparation and Analyses

The net amounts of carbon stored during the late maize cropping season (August–November) of each year were calculated using the method adapted from Tian et al. (2009):

$$\text{Soil OC sequestration} = \frac{\text{Increase in soil organic carbon stock due to treatments over control}}{\text{Number of experimental years}}$$

The evaluated carbon (mg/ha) was converted to CO_2 equivalents in Mg per hectare using the equation below:

$$CO_2 = C \times \frac{44\ \text{g/mole } CO_2}{12\ \text{g/mole C}}$$

For the pre- and post-cropped soil tests, representative soil samples to the depth of 0–15 cm in each plot were collected, air-dried and sieved through a 2-mm mesh size for analyses. Sieved soil, compost and biochar samples were analysed using standard methods (Page et al. 1982). The soil pH was determined in 1:1 soil–water suspension using a glass electrode pH metre. The macro-Kjeldahl method determined the total nitrogen of the soil, compost and biochar. Available phosphorus in the soil, compost and biochar was extracted using the Bray P1 method, and P in the extract was determined by a colorimeter. The organic carbon in soil, compost and biochar was determined using Walkley–Black wet oxidation method.

Bulk density was determined using a method described by Blake and Hartge (1986). The bulk density was estimated by inserting a fixed-volume steel ring (core sampler) into the soil at three randomly located points in each plot at 0–10 cm and 10–20 cm soil depths. Soil samples were removed from the core sampler, dried at 105 °C for 48 h and weighed. The volume of the soil was determined by calculating the volume of the core sampler used to extract the soil. The bulk density was expressed in a unit gram of oven-dry soil per cubic centimetre. Climatic data, particularly rainfall spread of the study area over a 5-year period, were collected from the Meteorological Station, Institute of Ecology and Environmental Studies, OAU, Ile-Ife, Nigeria.

2.3 Statistical Analysis

One-way analysis of variance (ANOVA) was used to test for statistical differences in the value of selected soil properties (pH, SOC, total N, available P, Ca^{2+}, CEC) and grain yield of maize over an a-five year period as influenced by the experimental variables using SAS 9.2 version at $p < 0.05$. The treatment means were separated at a 95% confidence limit using LSD.

3 Result and Discussion

3.1 Effect of Biochar and Compost Application on Soil pH

The pre-cropped soil properties of the experimental site are in Table 1. The soil pH was 6.35, an indication that the soil was slightly acidic. Generally, biochar is alkaline (Biederman and Harpole 2013; DeLuca et al. 2015; Shetty and Prakash 2020), as indicated in Table 2. In this study, biochar addition to the experimental field significantly increased soil pH, and this increase varied with different biochar addition and continuous cultivation of the field to maize.

Variations in the liming effects of biochar could be due to differences in their alkalinity nature (Adewole et al. 2020). Many authors, such as Peng et al. (2011), Ren

Table 1 Pre-cropped soil properties

Property	Value
pH (H_2O)	6.35
Organic carbon (g kg^{-1})	20.78
Total N (g kg^{-1})	1.94
Available P (mg kg^{-1})	16.39
Exchangeable cation (cmol kg^{-1})	
Ca^{2+}	4.10
Mg^{2+}	0.26
K^{+}	0.33
Na^{+}	0.46
CEC	5.15
Bulk density (g cm^{-3})	1.27
Soil texture	Sandy loam

Table 2 Selected properties of biochar and compost used

Property	Value					
	MCB	MSB	CDC	CDB	PTC	PTB
pH (H_2O)	10.77	11.10	8.83	9.67	9.13	9.27
Total carbon (g kg^{-1})	483.50	511.00	80.75	84.10	88.00	99.90
Total N (g kg^{-1})	10.90	12.10	2.48	2.10	2.16	2.36
C/N	44.35	42.23	32.56	40.05	40.74	42.33
Available P (mg kg^{-1})	70.00	65.35	39.48	78.96	35.29	108.57
Ash (%)	17.59	16.75	61.11	81.91	45.98	57.61

Adapted from Ilesanmi (2015) and Aina (2019)
MCB Maize cob biochar, *MSB* Maize stover biochar, *CDC* Cow dung compost, *CDB* Cow dung biochar, *PTC* Poultry droppings compost and *PTB* Poultry droppings biochar

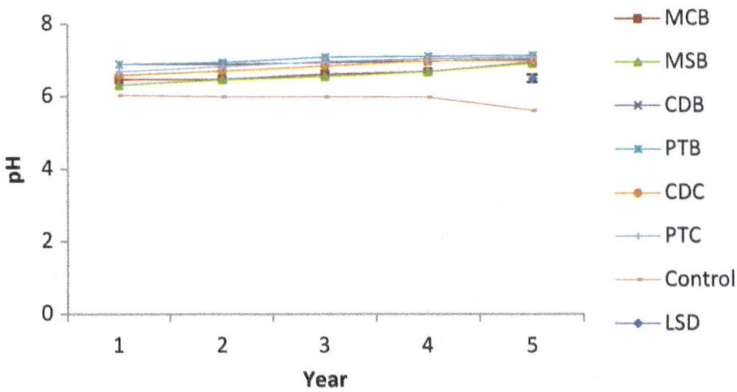

Fig. 2 Effect of biochar and compost application on soil pH. *MCB* Maize cob biochar, *MSB* Maize stover biochar, *CDC* Cow dung compost, *CDB* Cow dung biochar, *PTC* Poultry droppings compost and *PTB* Poultry droppings biochar

et al. (2019) and Adekiya et al. (2020), observed soil pH increase when biochar was applied to the soil. Plots with zero treatment application experienced reduced soil pH during the same period. African soils become more acidic as soils get depleted with cereal crop cultivation (Mahmood et al. 2017). There was a significant difference in soil pH between zero treatment application and other experimental variables (Fig. 2).

3.2 *Effect of Biochar and Compost Application on Soil Organic Carbon and Nitrogen*

The pre-cropped soil organic carbon (SOC) and total N were 20.78 and 1.94 g kg^{-1}, respectively. These values were not adequate as N is a major nutrient-demanding factor for the optimum production of maize (Schwalbert et al. 2018; Maresma

Galindo et al. 2019). There was a drastic reduction of SOC and total N after the first year of cultivation (Figs. 3 and 4). Apart from soil nutrients removal by the growing crop to enhance crop health during cultivation (Othman et al. 2020; Yang et al. 2020), native nutrients are generally reduced with native vegetation removal of tropical soils (Adesina 1998; Gay-des-Combes et al. 2017). Except for zero biochar application that had reduced SOC and total N during the yearly cultivation, other treatments increased significantly ($p < 0.05$) the values of SOC and total N. The inherent biochar's rich carbon content that has been studied by many scientists (Lefebvre et al. 2020; Tomczyk et al. 2020) accounted for this increase.

PTC/PTB treatment positively influenced SOC and total N during the 5 years of the field experiment. Variations in the type of feedstocks for biochar production could be responsible for this. The type of feedstock used and the conditions under which

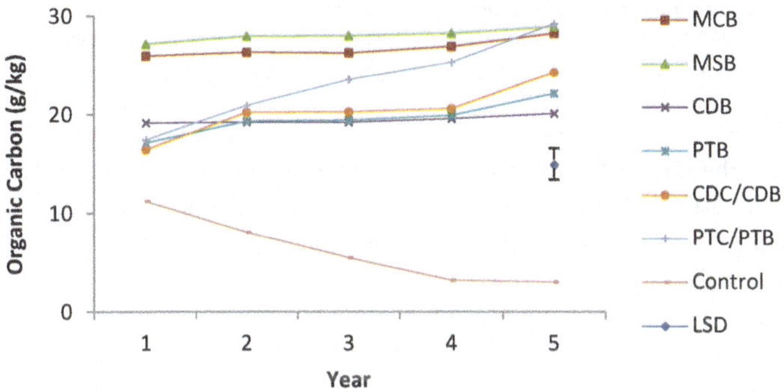

Fig. 3 Effect of biochar and compost application on soil organic carbon. *MCB* Maize cob biochar, *MSB* Maize stover biochar, *CDC* Cow dung compost, *CDB* Cow dung biochar, *PTC* Poultry droppings compost and *PTB* Poultry droppings biochar

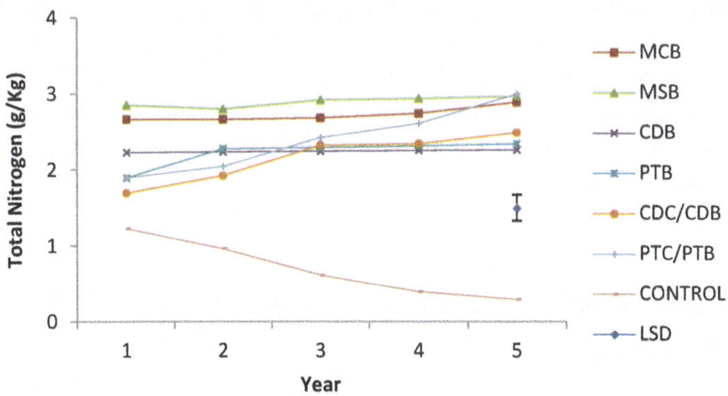

Fig. 4 Effect of biochar and compost application on total nitrogen of the soil. *MCB* Maize cob biochar, *MSB* Maize stover biochar, *CDC* Cow dung compost, *CDB* Cow dung biochar, *PTC* Poultry droppings compost and *PTB* Poultry droppings biochar

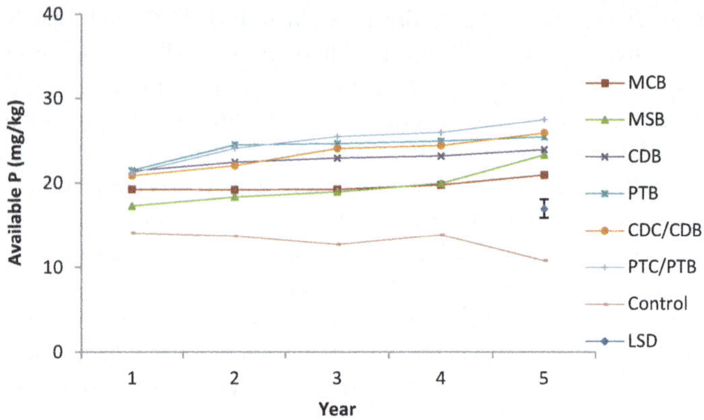

Fig. 5 Effect of biochar and compost application on available phosphorus of the soil. *MCB* Maize cob biochar, *MSB* Maize stover biochar, *CDC* Cow dung compost, *CDB* Cow dung biochar, *PTC* Poultry droppings compost and *PTB* Poultry droppings biochar

biochar is produced greatly affect its relative quantity and quality as a soil amendment (Ippolito et al. 2020). Biochar from animal waste demonstrated superiority over those from plant materials (Zhang et al. 2016). The long-term advantage of biochar in soil nutrient accessibility is due primarily to increased stability and slower release of nutrients from added organic matter (Joseph et al. 2015; Carrari et al. 2018).

3.3 *Effect of Biochar and Compost Application on Soil Available P*

Another positive influence of biochar application is the increase in available P (Fig. 5). Biochar and compost addition to soil enhanced P solubility and availability, whose inherent preference is a fixed state. In this study, significantly ($p < 0.05$), lowest value was obtained with zero treatment application compared to other variables. Lehmann et al. (2011) and Mensah and Frimpong (2018) observed increased available P, K and Zn when soil added biochar.

3.4 *Effect of Biochar and Compost Application on Soil Available Ca and CEC*

Crop cultivation without fertilizer application reduces inherent soil nutrients, including the CEC, due to the mining effect of the growing crop. However, biochar adsorption capability (Khan et al. 2018) enhanced the CEC, even with continuous maize cultivation (Figs. 6 and 7). This CEC was dominated by available Ca due to the presence of ash brought about by pyrolytic conditions (Yang et al. 2015). The CEC reduced by about 20% after the first maize cultivation and kept reducing to about 75% after the fifth maize cultivation under zero treatment application. These same parameters increased with biochar and compost application under continuous maize cultivation. Adding biochar with compost demonstrated better retention capability due to the increased ability to adsorb available cations (Agegnehu et al. 2017), increased surface area and great charge density (Liang et al. 2006) than the native soil organic matter alone. In this study, PTC/PTB treatment demonstrated the highest adsorption capacity. Differences in the feedstocks used for biochar production must have accounted for this (Karbout et al. 2019).

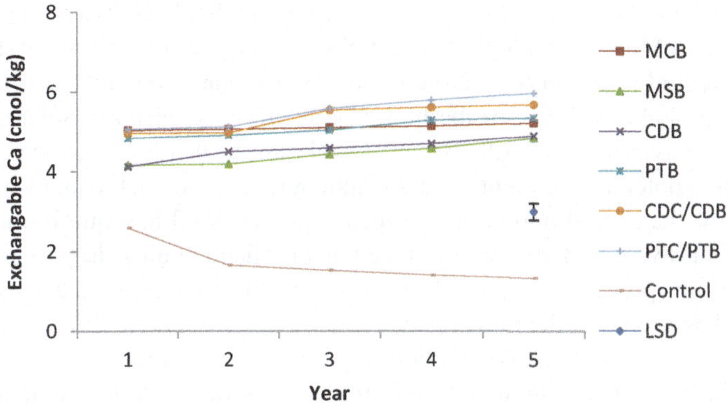

Fig. 6 Effect of biochar and compost application on available calcium of the soil. *MCB* Maize cob biochar, *MSB* Maize stover biochar, *CDC* Cow dung compost, *CDB* Cow dung biochar, *PTC* Poultry droppings compost and *PTB* Poultry droppings biochar

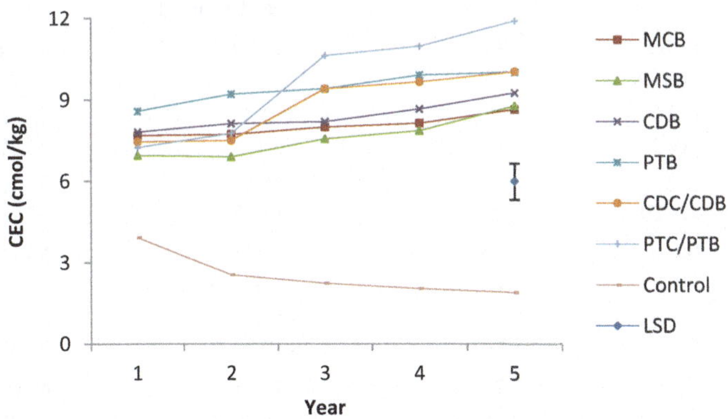

Fig. 7 Effect of biochar and compost application on soil cation exchange capacity. *MCB* Maize cob biochar, *MSB* Maize stover biochar, *CDC* Cow dung compost, *CDB* Cow dung biochar, *PTC* Poultry droppings compost and *PTB* Poultry droppings biochar

3.5 *Effect of Biochar and Compost Application on Soil Carbon Sequestration*

The effects of biochar application and soil depth on soil carbon sequestration are presented in Table 3. In this study, the co-application of poultry droppings compost and biochar (PTC/PTB) had the highest potential for C sequestration compared to other amendments. Also, more C was sequestered in the topsoil, 0–10 cm of the soil, than in the subsoil, 10–20 cm. The PTC/PTB treatment sequestered 18.97 and 11.46 t C ha^{-1} $year^{-1}$ on the topsoil and subsoil, respectively. However, MCB applied sequestered the least, 12.39 and 2.20 t C ha^{-1} $year^{-1}$ on the topsoil and subsoil, respectively.

Of the two biochar from plant feedstock (MCB, MSB), MSB, with a much higher C content than MCB, had a higher potential for C sequestration. Also, PTB had the highest potential for C sequestration among the biochar and compost from animal feedstock, probably because of the high C content in the biochar. Sukartono et al. (2011), Purakayastha et al. (2015) and Qambrani et al. (2017) observed high C sequestration potentials brought about by high SOC gains when biochar was applied. In this study, the overall optimal performance of PTC/PTB in sequestering C could be due to improved reactive site introduced by the added poultry droppings compost and poultry droppings biochar. Biochar addition to soil increased the recalcitrant soil C pool and reduced the rate of accumulation of atmospheric CO_2 concentration (Smith 2016). In this study, soil carbon sequestered was directly related to the rate of atmospheric CO_2 accumulated. Variation in C sequestration potentials among

Table 3 Effect of biochar application and soil depth on soil carbon sequestration rate

Treatment	Soil depth (cm)	Bulk density (g cm^{-3})	Soil carbon sequestration rate (t C ha^{-1} $year^{-1}$)	t CO_2 (t ha^{-1})
MCB	0–10	1.33	12.39	45.44
	10–20	1.35	2.20	8.07
MSB	0–10	1.32	12.45	45.66
	10–20	1.33	6.13	22.47
CDB	0–10	1.38	15.62	57.33
	10–20	1.40	14.63	53.73
PTB	0–10	1.35	14.21	52.15
	10–20	1.40	9.88	13.23
CDC/CDB	0–10	1.36	18.27	67.05
	10–20	1.38	10.53	38.65
PTC/PTB	0–10	1.33	18.97	69.62
	10–20	1.34	11.46	27.38
Control	0–10	1.29	–	–
	10–20	1.31	–	–

MCB Maize cob biochar, *MSB* Maize stover biochar, *CDC* Cow dung compost, *CDB* Cow dung biochar, *PTC* Poultry droppings compost and *PTB* Poultry droppings biochars biochar and compost from different feedstock may not be unconnected with different pyrolysis temperatures, stability potential and Chemistry of native C (Peng et al. 2011).

3.6 *Effect of Biochar and Compost Application on the Dry Stover and Grain Yield of Maize*

The effects of biochar application on the dry stover and grain yield of maize are presented in Figs. 8 and 9. Chemical changes in soil after biochar and compost application reflect the properties of the added amendment (Trupiano et al. 2017), which invariably translated to the variations in the dry stover and grain yields of maize obtained. Except for the first year of the experiment, biochar and compost application resulted in gradual dry stover and grain yield increase with continuous maize crop cultivation. Fagbenro et al. (2018) obtained increased total dry matter yield of maize when maize crop residue biochar and uncharred poultry manure were compared. Also, Ndor et al. (2018) obtained optimal growth response and yield of improved variety of soybean (TGX 1448-E2) with 10 t ha^{-1} of biochar application.

In this study, therefore, increased soil organic carbon and water holding capacity brought about by the applied biochar and compost must have accounted for a gradual increase in dry stover and grain yields of maize, even with continuous cultivation. In the plots without biochar and compost, a sharp reduction began from 2nd year of cultivation, and about 75% reduction in dry stover and grain yields of maize was obtained for the 5th cultivation. This observation justifies the need to appropriately replenish the soil of this type for maize cultivation.

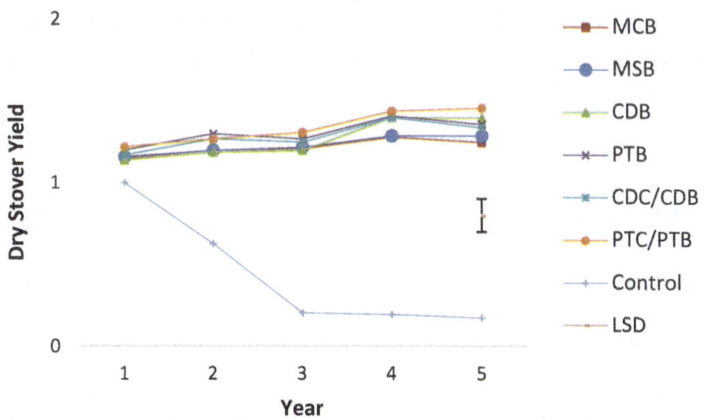

Fig. 8 Effect of biochar and compost application on the dry stover yield (t ha^{-1}) of maize. *MCB* Maize cob biochar, *MSB* Maize stover biochar, *CDC* Cow dung compost, *CDB* Cow dung biochar, *PTC* Poultry droppings compost and *PTB* Poultry droppings biochar

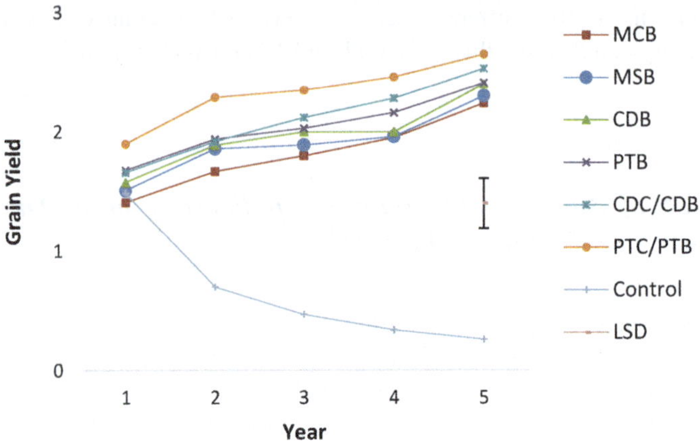

Fig. 9 Effect of biochar and compost application on the grain yield (t ha^{-1}) of maize. *MCB* Maize cob biochar, *MSB* Maize stover biochar, *CDC* Cow dung compost, *CDB* Cow dung biochar, *PTC* Poultry droppings compost and *PTB* Poultry droppings biochar

4 Conclusion

When used as soil amendments, biochar and compost obtained from different degradable plant and animal farm wastes had different effects on the growth and yield of maize, soil chemical properties and sequestered carbon in the soil. With continuous maize cultivation over 5 years, most soil properties (organic carbon, total nitrogen, phosphorus and cation exchangeable capacity) increased, while soil pH remained relatively constant when biochar and compost, either singly or in combination, were used as soil amendments, but these properties decreased in zero treatment plots. We also concluded that poultry droppings biochar with compost in equal proportion sequestered the highest carbon on the topsoil than the subsoil when compared with other treatment applications.

References

Adekiya AO, Olaniran AF, Adenusi TT, Aremu C, Ejue WS, Iranloye YM, Gbadamosi A, Olayanju A (2020) Effects of cow dung and wood biochars and green manure on soil fertility and tiger nut (*Cyperus esculentus* L.) performance on a savanna Alfisol. Scientific Reports 10(1):1–10

Adesina FA (1998) Plant species characteristics and vegetation dynamics in the tropics. Int J Environ Stud 33:67–78

Adewole MB, Ilesanmi AO, Odesola IF, Isichei AO (2020) Effect of biochar application rates on soil properties, growth and yield of maize under greenhouse conditions. African Journal of Environmental Health Sciences. 7:125–135

Aina BM (2019) Influence of compost fortified with biochar on carbon sequestration, growth and yield of maize in Ile-Ife, Sothwestern Nigeria. A PhD thesis submitted to the Postgraduate College, Obafemi Awolowo University, Ile-Ife, Nigeria, 185 pp

Agegnehu G, Bass AM, Nelson PN, Bird MI (2016) Benefits of biochar, compost and biochar-compost for soil quality, maize yield and greenhouse gas emissions in a tropical agricultural soil. Science and Total Environment 543:295–306

Agegnehu G, Srivastava AK, Bird MI (2017) The role biochar and biochar-composts in improving soil quality and crop performance: a review. Appl Soil Ecol 119:156–170

Akmal M, Maqbool Z, Khan KS, Hussain Q, Ijaz SS, Iqbal M, Aziz I, Hussain A, Abbas MS, Rafa HU (2019) Integrated use of biochar and compost to improve soil microbial activity, nutrient availability, and plant growth in arid soil. Arab J Geosci 12:232

Alabi DM, Akintola OM, Famakinwa M (2018) Perception of maize processor towards utilization of maize value addition techniques: implications for rural entrepreneurship development—a study in Nigeria. Journal of Agricultural Sciences 13(2):141–152

Biederman LA, Harpole WS (2013) Biochar and its effects on plant productivity and nutrient cycling: a meta-analysis. GCB Bioenergy 5(2):202–214

Blake GR, Hartge KH (1986) Particle density. Methods of Soil Analysis: Part 1 Physical and Mineralogical Methods 5:377–382

Carrari E, Ampoorter E, Bussotti F, Coppi A, Nogales AG, Pollastrini M, Verheyen K, Selvi F (2018) Effects of charcoal hearth soil on forest regeneration: evidence from a two-year experiment on tree seedlings. For Ecol Manage 427:37–44

Chen Y, Camps-Arbestain M, Shen Q, Singh B, Cayuela ML (2018) The long-term role of organic amendments in building soil nutrient fertility: a meta-analysis and review. Nutrient Cycling in Agro-Ecosystems 111(2–3):103–125

DeLuca TH, Gundale MJ, MacKenzie MD, Jones DL (2015) Biochar effects on soil nutrient transformations. Biochar for Environmental Management: Science, Technology and Implementation 2:421–454

Dunham RA, Elaswad A (2018) Catfish biology and farming. Annual Review of Animal Biosciences 6:305–325

El-Naggar A, El-Naggar AH, Shaheen SM, Sarkar B, Chang SX, Tsang DC, Rinklebe J, Ok YS (2019) Biochar composition-dependent impacts on soil nutrient release, carbon mineralization, and potential environmental risk: a review. J Environ Manage 241:458–467

Elum ZA, Momodu AS (2017) Climate change mitigation and renewable energy for sustainable development in Nigeria: a discourse approach. Renew Sustain Energy Rev 76:72–80

Fagbenro JA, Oshunsanya SO, Aliku O, Odekunle D (2018) Effects of maize crop residue, uncharred poultry manure and inorganic fertilizer on maize grown in pedogenic horizons of a tropical Alfisol. BiocharTec Journal 1:15–27

Gay-des-Combes JM, Sanz Carrillo C, Robroek BJM, Jassey VEJ, Mills RTE, Arif MS, Falquet L, Frossard E, Buttler A (2017) Tropical soils degraded by slash-and-burn cultivation can be recultivated when amended with ashes and compost. Ecol Evol 7(14):5378–5388

Hardy B, Cornelis JT, Houben D, Leifeld J (2016) Evaluation of the long-term effect of biochar on properties of temperate agricultural soil at pre-industrial charcoal kiln sites in Wallonia, Belgium. Eur J Soil Sci 6:1–10

Ilesanmi AO (2015) Effects of different sources and proportions of biochar on soil carbon sequestration rates and yield of maize (*Zea mays* L.). A PhD thesis submitted to the Postgraduate College, Obafemi Awolowo University, Ile-Ife, Nigeria, 200 pp

Ippolito JA, Cui L, Kammann C, Wrage-Monnig N, Estavillo JM, Fuertes T, Cayuela ML, Sigua G, Novak J, Spokas K, Borchard N (2020) Feedstock choice, pyrolysis temperature and type influence biochar characteristic: a comprehensive meta-data analysis review. Biochar 2:421–438

Jindo K, Mizumoto H, Sawada Y, Sanchez-Monedero M, Sonoki T (2014) Physical and chemical characterization of biochars derived from different agricultural residues. Biogeosciences 11:6613–6621

Joseph S, Doug P, Dawson K, Mitchell DR, Rawal A, James H, Taherymoosavi S, Van Zwieten L, Joshua R, Donne S (2015) Feeding biochar to cows: an innovative solution for improving soil fertility and farm productivity. Pedosphere 25:666–679

Karbout N, Bol R, Brahim N, Moussa M, Bousnina H (2019) Applying biochar from date palm waste residues to improve the organic matter, nutrient status and water retention in sandy oasis soils. Journal of Research in Environmental and Earth Sciences 7:203–209

Khan TF, Salma MU, Hossain SA (2018) Impacts of different sources of biochar on plant growth characteristics. Am J Plant Sci 9:1922–1934

Kushwah S, Dhir A, Sagar M (2019) Understanding consumer resistance to the consumption of organic food. A study of ethical consumption, purchasing, and choice behaviour. Food Qual Prefer 77:1–14

Lal R (2004) Soil carbon sequestration to mitigate climate change. Geoderma 123:1–22

Lal R (2019) Conceptual basis of managing soil carbon inspired by nature and driven by science. J Soil Water Conserv 74(2):29–34

Lefebvre D, Williams A, Meersmans J, Kirk GJ, Sohi S, Goglio P, Smith P (2020) Modelling the potential for soil carbon sequestration using biochar from sugarcane residues in Brazil. Nature 10:1–11

Lehmann J, Rillig MC, Thies J, Masiello CA, Hockaday WC, Crowley D (2011) Biochar effects on soil biota–a review. Soil Biol Biochem 43(9):1812–1836

Leno N (2020) Soil carbon sequestration: basis and basics. Book Publisher International, India, p 53

Liang B, Lehmann J, Solomon D, Kinyang J, Grossman J, O'Neill B, Skjemstad J, Thies J, Luizão FJ, Petersen J, Neves EG (2006) Black carbon increase cation exchangeable capacity in soil. Soil Sci 70:1719–1730

Liu XF, Han W, Liu Z, Wang X, Zhao X (2016) Impacts of biochar amended soils on ryegrass (*Lolium perenne*) growth under different water stress conditions. Int J Agric Biol 18:630–636

Macdonald LM, Farrell M, Krull ES (2014) Plant growth response to biochar addition: an Australian soils perspective. Biol Fertil Soils 50:169–178

Mahmood F, Khan I, Ashraf U, Shahzad T, Hussain S, Shahid M, Abid M, Ullah S (2017) Effects of organic and inorganic manures on maize and their residual impact on soil physico-chemical properties. J Soil Sci Plant Nutr 17(1):22–32

Maresma Galindo Á, Martínez Casasnovas JA, Santiveri Morata P, Lloveras Vilamanyà J (2019) Nitrogen management in double-annual cropping system (barley-maize) under irrigated Mediterranean environments. Eur J Agron 103:98–107

Mensah AK, Frimpong KA (2018) Biochar and or compost applications improve soil properties, growth, and yield of maize grown in acidic rainforest and coastal Savannah soils in Ghana. International Journal of Agronomy 7:1–8

Monday SA (2018) Economic value of maize (*Zea mays* L.) in Nigeria and its impacts on the global food production. International Journal of Scientific World 6(1):27–30

Ndor E, Jayeoba OJ, Dauda SN (2018) Response of soybean (*Glycine max* L. Merrill) varieties to biochar soil amendment in Southern Guinea savanna of Nigeria. BiocharTec Journal 1:44–51

Nunes LJ, Meireles CI, Gomes CJ, Almeida-Ribeiro NM (2020) Forest contribution to climate change mitigation: management oriented to carbon capture and storage. Climate 8(21):1–20

Odesola IF, Owoseni TA (2010) Development of local technology for small-scale biochar production processes from agricultural wastes. Journal of Emerging Trends in Engineering and Applied Sciences (JETEAS) 1(2):205–208

Ogunniyan DJ, Ojo DK, Olakojo SA, Talabi OA (2019) Diallel analysis of maize inbred lines for agronomic traits in nitrogen stress and optimal conditions. Ghana Journal of Agricultural Science 54(1):10–23

Oladejo JA, Adetunji MO (2012) Economic analysis of maize (*Zea mays* L.) production in Oyo State of Nigeria. Agricultural Science Research Journals 2(2):77–83

Olmo M, Lozano AM, Barron V, Villar R (2016) Spatial heterogeneity of soil biochar content affects soil quality and wheat growth and yield. Sci Total Environ 562:690–700

Olson K (2013) Soil organic carbon sequestration, storage, retention and loss in U.S. croplands: issues paper for protocol development. Geoderma 195–196:201–206

Othman YA, Tahat MM, Alananbeh KM, Leskovar D (2020) Soil health and sustainable agriculture. Sustainability 12:4859–4885

Page AL, Miller RH, Keeney DR (1982) Methods of soil analysis, part 2, chemical and microbiological properties. American Society of Agronomy Inc., Madison, WI, USA, p 556

Peng XYLL, Ye LL, Wang CH, Zhou H, Sun B (2011) Temperature-and duration-dependent rice straw-derived biochar: Characteristics and its effects on soil properties of an Ultisol in southern China. Soil and Tillage Research 112(2):159–166

Purakayastha TJ, Kumari S, Pathak H (2015) Characterisation, stability, and microbial effects of four biochars produced from crop residues. Geoderma 239:293–303

Qambrani NA, Rahman MM, Won S, Shim S, Ra C (2017) Biochar properties and eco-friendly applications for climate change mitigation, waste management, and wastewater treatment: a review. Renew Sustain Energy Rev 79:255–273

Ren H, Lv C, Fernández-García V, Huang B, Yao J, Ding W (2019) Biochar and PGPR amendments influence soil enzyme activities and nutrient concentrations in a eucalyptus seedling plantation. Biomass Conversion and Biorefinery 1–10

Schwalbert R, Amado TJ, Horbe TA, Stefanello LO, Assefa Y, Prasad PV, Rice CW, Ciampitti IA (2018) Corn yield response to plant density and nitrogen: spatial models and yield distribution. Agron J 110(3):970–982

Shetty R, Prakash NB (2020) Effect of different biochars on acid soil and growth parameters of rice plants under aluminium toxicity. Sci Rep 10(1):1–10

Singh SV, Dhyani VC, Chaturvedi S, Datta D (2019) Biochar: an eco-friendly residue management approach. Indian Farming 69(8):27–29

Smith P (2016) Soil carbon sequestration and biochar as negative emission technologies. Glob Change Biol 22:1315–1324

Sukartono WH, Nugroho WH, Kusuma Z (2011) Simple biochar production generated from cattle dung and coconut shell. Journal of Basic and Applied Sciences Research 1(10):1680–1685

Tian G, Granato TC, Cox AE, Pietz RI, Carlson CR Jr, Abedin Z (2009) Soil carbon sequestration resulting from long-term application of biosolids for land reclamation. J Environ Qual 38(1):61–74

Tomczyk A, Sokolowska A, Boguta P (2020) Biochar physicochemical properties: pyrolysis temperature and feedstock kind effects. Revised Environmental Science and Biotechnology 3:1–25

Trupiano D, Cocozza C, Baronti S, Amendola C, Vaccari FP, Lustrato G, Lonardo S, Fantasma F, Tognetti R, Scippa GS (2017) The effects of biochar and its combination with compost on lettuce (*Lactuca sativa* L.) growth, soil properties, and soil microbial activity and abundance. International Journal of Agronomy 21:1–12

Yang F, Lee X, Wang B (2015) Characterization of biochars produced from seven biomasses grown in three different climate zones. Chin J Geochem 34:592–600

Yang X, Meng J, Lan Y, Chen W, Yang T, Yuan J, Liu S, Han J (2017) Effects of maize stover and its biochar on soil CO_2 emissions and labile organic carbon fractions in northeast China. Agricultural Ecosystem and Environment 240:24–31

Yang T, Siddique KH, Liu K (2020) Cropping systems in agriculture and their impact on soil health—a review. Global Ecology and Conservation 23:1–13

Zhang J, Chen G, Sun H, Zhou S, Zou G (2016) Straw biochar hastens organic matter degradation and produces nutrient-rich compost. Biores Technol 200:876–883

The manufacturer's authorised representative in the EU is Springer Nature Customer Service Centre GmbH, Europaplatz 3, 69115 Heidelberg, Germany. If you have any concerns regarding our products, please contact ProductSafety@springernature.com

Printed and bound by CPI Group (UK) Ltd, Croydon, CR0 4YY
28/09/2026
02224075-0018